67 PRIZEWINNING PLYWOOD PROJECTS

Alfred W. Lees

 Sterling Publishing Co., Inc.   New York

1   3   5   7   9   10   8   6   4   2

First published in paperback in 1987 by Sterling Publishing Co., Inc.
Two Park Avenue, New York, N.Y. 10016
Originally published in hardcover © 1984 by Grolier Book Clubs, Inc.
Distributed in Canada by Oak Tree Press Ltd.
℅ Canadian Manda Group, P.O. Box 920, Station U
Toronto, Ontario, Canada M8Z 5P9
Distributed in the United Kingdom by Blandford Press
Link House, West Street, Poole, Dorset BH15 1LL, England
Distributed in Australia by Capricorn Ltd.
P.O. Box 665, Lane Cove, NSW 2066
Manufactured in the United States of America
All rights reserved
Library of Congress Catalog Card No.: 84-23742
Sterling ISBN 0-8069-6520-7 Paper

CONTENTS – INDEX

INDEX OF PROJECTS: ALPHABETICAL LISTING

INDEX: ALPHABETICAL LISTING

INDEX OF PROJECTS: QUANTITY OF PLYWOOD ═══════

INDEX: QUANTITY LISTING

INTRODUCTION

There are many reasons for writing a book: To "set the record straight" (as with political memoirs); to clarify or revise historical perspective—or to leave one's mark on it (as with an autobiography); simply to bolster one's ego (as with "vanity press" publications); to instruct or entertain. Perhaps the happiest motivation is when the book is a "command performance," specifically asked for by potential readers.

That, I'm glad to say, is the reason for this book. Throughout the many years that my annual plywood projects contest has been published in POPULAR SCIENCE magazine, readers have persistently asked that all the projects be gathered into a book—especially since many of the plans for prizewinners in the early years have long passed out of print.

But such a book has not been easy to prepare. When I publish the annual portfolio of winning projects in each August issue of the magazine, I don't have space to include construction plans or step-by-step instructions for assembly. I simply show each project in a color photo (or two, if there's a dual use to demonstrate) and include a panel layout to show how all the plywood parts were cut from one or more sheets (since efficient utilization of the plywood is an important condition for qualifying). Experienced home-workshoppers can often build the project from the layout diagram alone, but most woodworkers prefer to work from a detailed plan that shows how the parts go together. It's just such a plan that the American Plywood Association (APA)—which has worked with me on the contest from its inception—prepares for all (or nearly all) of the prizewinning projects. And these are the plans that are no longer available for many of the early projects.

So I asked APA's permission to reproduce all the plans from my seven years' file of them. But even that did not create a book, because most of these plan folders lacked verbal instructions. Ultimately I had to restudy each of the 67 projects and all the APA plans so I could write a how-to text for each. In a few cases, no plans folder had ever been prepared, so there I had to start from scratch.

The chore was lightened by the fact that it was rather like attending a class reunion to reacquaint myself with old, half-forgotten friends. I've come away from this seven-month-long reunion with renewed pride in my association with the contest. I'd say APA and I have been doing right well by it, and my faith in the shop skills and design abilities of POPULAR SCIENCE readers has been newly rewarded.

This seemed an ideal time to collect the 67 prize-winning *plywood* projects of the first seven years of the contest into an organized, permanent plans portfolio—and here it is. I've divided those projects into two basic categories—Indoor and Outdoor—because *where* you'll use the project influences the type of plywood you'll buy for it (as well as the type of glue and fasteners you buy to assemble it, and the type of finish you'll apply). Then, within these categories, I've grouped the plans by *function*, so that if, for example, you want to build a desk, you can turn to the section where all of the prizewinning desks are grouped, and select the best design for your particular needs. Just check the contents page to find where each grouping begins.

One general note on the plans themselves: Sometimes, where entrants took special pains to allow for saw-kerf waste in their layout dimensions (usually ⅛″ per cut), we've retained these specifications. Often, though, we've simplified the diagrams by assuming you'd center your cut on each layout line. This is important where mating parts are concerned. When cutting notches to take panel thicknesses, they should be slightly oversize to permit a slip fit—especially after pieces are coated with paint and are intended for disassembly (for storage or portability).

My book begins with two chapters on plywood itself—how to buy it, and how to work with it. Much of this, many readers will already know from experience, but the facts are there for those who are less familiar with this versatile and relatively inexpensive material.

Inundated in entries, APA's Maryann Olson and PS Editor Alfred W. Lees tackle their annual judging.

But how did the whole concept of this now-famous contest begin? It dates all the way back to this nation's Bicentennial, in 1976. Four years earlier, POPULAR SCIENCE had itself celebrated its 100th Anniversary. Perhaps inspired by revolutionary fervor, the then-editor-in-chief (long since retired)—out of that infinite wisdom that automatically accrues to such eminence—abruptly decided that henceforth there would be no shop projects in POPULAR SCIENCE. All of the magazine's how-to pages would be, from that moment, devoted exclusively to home improvement: built-ins, add-ons, upgrades. The rule of thumb came to be "If you can pick it up after you've built it, it's not for PS."

I tend to be conservative by nature, doubting that much radical change *since* the American Revolution has been true progress—and this was no exception. So when, by chance, at this crucial moment in the magazine's history, the Tacoma office of the APA phoned me to announce that POPULAR SCIENCE had been chosen as the magazine they most wanted to work with on an annual projects contest, I seized the opportunity. I quickly established the basic guidelines and editorial scheduling and, to my relief, won approval from PS management. This was my way of assuring that—once a year, at least—the very *best* projects PS readers could produce would assume their proper place in our pages.

So you might say there was a degree of subversiveness in the original concept of this contest. But who dares argue with the success it has been in the years since?

I can't close these comments without expressing sincere thanks to the people without whose help this book wouldn't be in your hands (and workshops) today. To John Sill, publisher of the PS Book Division, who first proposed the project to me; to book editor Irna Gadd, whose patience is matched only by her skills; to artist Carl De Groote, whose supreme draftsmanship has helped bring each August's portfolio of prizewinners to life (and a sprinkling of whose work appears through these pages); and, of course, to the Information Services Division of Tacoma's APA—to its genial head, Dave Rogoway, and to his indefatigable Program Coordinator, Maryann Olson. No one knows better than Maryann herself what she's contributed to the success of the contest—and of this book. Finally, it goes without saying that the major contribution to this book remains the projects themselves. Without the design ingenuity and tool skills of our contestants, there would be no annual project portfolio, and thus no book. To every prizewinner in this volume—and to the several thousand POPULAR SCIENCE readers who've taken the pains to enter this contest through the years—my personal thanks.

ALFRED W. LEES
Reader Activities Editor
POPULAR SCIENCE

WEIGHTS AND MEASURES

UNIT	ABBREVIATION	EQUIVALENTS IN OTHER UNITS OF SAME SYSTEM	METRIC EQUIVALENT
Weight			
Avoirdupois			
ton			
short ton		20 short hundredweight, 2000 pounds	0.907 metric tons
long ton		20 long hundredweight, 2240 pounds	1.016 metric tons
hundredweight	cwt		
short hundredweight		100 pounds, 0.05 short tons	45.359 kilograms
long hundredweight		112 pounds, 0.05 long tons	50.802 kilograms
pound	lb *or* lb av *also* #	16 ounces, 7000 grains	0.453 kilograms
ounce	oz *or* oz av	16 drams, 437.5 grains	28.349 grams
dram	dr *or* dr av	27.343 grains, 0.0625 ounces	1.771 grams
grain	gr	0.036 drams, 0.002285 ounces	0.0648 grams
Troy			
pound	lb t	12 ounces, 240 pennyweight, 5760 grains	0.373 kilograms
ounce	oz t	20 pennyweight, 480 grains	31.103 grams
pennyweight	dwt *also* pwt	24 grains, 0.05 ounces	1.555 grams
grain	gr	0.042 pennyweight, 0.002083 ounces	0.0648 grams
Apothecaries'			
pound	lb ap	12 ounces, 5760 grains	0.373 kilograms
ounce	oz ap	8 drams, 480 grains	31.103 grams
dram	dr ap	3 scruples, 60 grains	3.887 grams
scruple	s ap	20 grains, 0.333 drams	1.295 grams
grain	gr	0.05 scruples, 0.002083 ounces, 0.0166 drams	0.0648 grams
Capacity			
U.S. Liquid Measure			
gallon	gal	4 quarts (2.31 cubic inches)	3.785 litres
quart	qt	2 pints (57.75 cubic inches)	0.946 litres
pint	pt	4 gills (28.875 cubic inches)	0.473 litres
gill	gi	4 fluidounces (7.218 cubic inches)	118.291 millilitres
fluidounce	fl oz	8 fluidrams (1.804 cubic inches)	29.573 millilitres
fluidram	fl dr	60 minims (0.225 cubic inches)	3.696 millilitres
minim	min	1/60 fluidram (0.003759 cubic inches)	0.061610 millilitres
U.S. Dry Measure			
bushel	bu	4 pecks (2150.42 cubic inches)	35.238 litres
peck	pk	8 quarts (537.605 cubic inches)	8.809 litres
quart	qt	2 pints (67.200 cubic inches)	1.101 litres
pint	pt	½ quart (33.600 cubic inches)	0.550 litres
British Imperial Liquid and Dry Measure			
bushel	bu	4 pecks (2219.36 cubic inches)	0.036 cubic metres
peck	pk	2 gallons (554.84 cubic inches)	0.009 cubic metres
gallon	gal	4 quarts (277.420 cubic inches)	4.545 litres
quart	qt	2 pints (69.355 cubic inches)	1.136 litres
pint	pt	4 gills (34.678 cubic inches)	568.26 cubic centimetres
gill	gi	5 fluidounces (8.669 cubic inches)	142.066 cubic centimetres
fluidounce	fl oz	8 fluidrams (1.7339 cubic inches)	28.416 cubic centimetres
fluidram	fl dr	60 minims (0.216734 cubic inches)	3.5516 cubic centimetres
minim	min	1/60 fluidram (0.003612 cubic inches)	0.059194 cubic centimetres
Length			
mile	mi	5280 feet, 320 rods, 1760 yards	1.609 kilometres
rod	rd	5.50 yards, 16.5 feet	5.029 metres
yard	yd	3 feet, 36 inches	0.914 metres
foot	ft *or* '	12 inches, 0.333 yards	30.480 centimetres
inch	in *or* "	0.083 feet, 0.027 yards	2.540 centimetres
Area			
square mile	sq mi *or* m²	640 acres, 102,400 square rods	2.590 square kilometres
acre		4840 square yards, 43,560 square feet	0.405 hectares, 4047 square metres
square rod	sq rd *or* rd²	30.25 square yards, 0.006 acres	25.293 square metres
square yard	sq yd *or* yd²	1296 square inches, 9 square feet	0.836 square metres
square foot	sq ft *or* ft²	144 square inches, 0.111 square yards	0.093 square metres
square inch	sq in *or* in²	0.007 square feet, 0.00077 square yards	6.451 square centimetres
Volume			
cubic yard	cu yd *or* yd³	27 cubic feet, 46,656 cubic inches	0.765 cubic metres
cubic foot	cu ft *or* ft³	1728 cubic inches, 0.0370 cubic yards	0.028 cubic metres
cubic inch	cu in *or* in³	0.00058 cubic feet, 0.000021 cubic yards	16.387 cubic centimetres

METRIC SYSTEM

UNIT	ABBREVIATION		APPROXIMATE U.S. EQUIVALENT		
		Length			
		Number of Metres			
myriametre	mym	10,000	6.2 miles		
kilometre	km	1000	0.62 mile		
hectometre	hm	100	109.36 yards		
dekametre	dam	10	32.81 feet		
metre	m	1	39.37 inches		
decimetre	dm	0.1	3.94 inches		
centimetre	cm	0.01	0.39 inch		
millimetre	mm	0.001	0.04 inch		
		Area			
		Number of Square Metres			
square kilometre	sq km *or* km²	1,000,000	0.3861 square miles		
hectare	ha	10,000	2.47 acres		
are	a	100	119.60 square yards		
centare	ca	1	10.76 square feet		
square centimetre	sq cm *or* cm²	0.0001	0.155 square inch		
		Volume			
		Number of Cubic Metres			
dekastere	das	10	13.10 cubic yards		
stere	s	1	1.31 cubic yards		
decistere	ds	0.10	3.53 cubic feet		
cubic centimetre	cu cm *or* cm³ *also* cc	0.000001	0.061 cubic inch		
		Capacity			
		Number of Litres	*Cubic*	*Dry*	*Liquid*
kilolitre	kl	1000	1.31 cubic yards		
hectolitre	hl	100	3.53 cubic feet	2.84 bushels	
dekalitre	dal	10	0.35 cubic foot	1.14 pecks	2.64 gallons
litre	l	1	61.02 cubic inches	0.908 quart	1.057 quarts
decilitre	dl	0.10	6.1 cubic inches	0.18 pint	0.21 pint
centilitre	cl	0.01	0.6 cubic inch		0.338 fluidounce
millilitre	ml	0.001	0.06 cubic inch		0.27 fluidram
		Mass and Weight			
		Number of Grams			
metric ton	MT *or* t	1,000,000	1.1 tons		
quintal	q	100,000	220.46 pounds		
kilogram	kg	1,000	2.2046 pounds		
hectogram	hg	100	3.527 ounces		
dekagram	dag	10	0.353 ounce		
gram	g *or* gm	1	0.035 ounce		
decigram	dg	0.10	1.543 grains		
centigram	cg	0.01	0.154 grain		
milligram	mg	0.001	0.015 grain		

HOW TO CHOOSE PLYWOOD PANELS

Plywood is a versatile and dependable project material—as long as you select the right type and grade for your intended use. It has certain advantages over solid lumber, the major one being that it comes in four-foot-wide panels, sparing you the chore of edge-gluing several planks when you need a wide piece, as for a tabletop or broad shelf.

Since plywood is fabricated with cross-laminated veneers, it is more resistant to warping and cupping than most lumber; it's well known for its dimensional stability—it shrinks and swells less than solid wood. It won't split or crumble, and the water-proof-glue bond in good-quality exterior plywood won't separate, even when exposed to extreme moisture conditions.

Which is the best plywood for your project? As a general rule, if you'll be using your project outdoors, or in an area of high humidity (a bathroom or kitchen, for example) choose an exterior type. Made with high-quality veneers and bonded with a water-proof glue, exterior plywood is the most durable type, and least likely to delaminate. Take special care to fill, prime, and seal any edges that are exposed to weather (especially edges in contact with damp earth or patios, such as the base of a piece of plywood furniture). For some projects (such as the boats in this book) you may want to ask your dealer to order marine plywood—a premium-grade exterior panel.

For most indoor projects, there's no need to pay extra for exterior quality. Though indoor types are often made with the same

GRADE STAMP

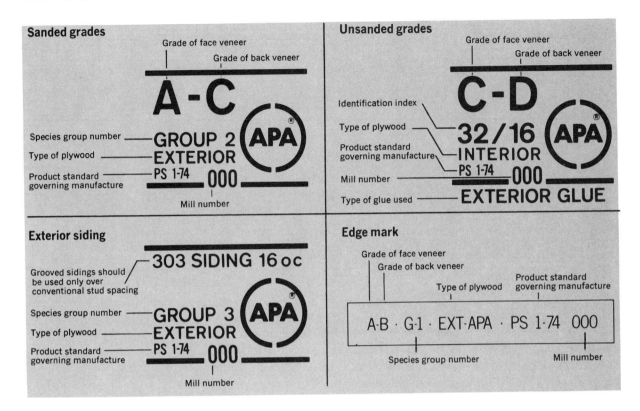

waterproof glue as exterior panels (you'll note that many interior-grade stamps will specify "exterior glue"), interior plywood may have lower-quality core veneers, so it should never be substituted for exterior grades.

As I've indicated, the grade of plywood is determined by the quality of its veneers. The two outer veneers are what you're concerned about, and they are classified by grading rules that are standard throughout the industry. Veneers are rated from A to D, according to the number of knots and patched flaws or voids in each. These grades are the most prominent part of any American Plywood Association grade trademark, as shown in the sample on page 3. Such trademarks are stamped on the *back* face and/or edge of all softwood panels. The first letter refers to the veneer on the front face of the panel; the second letter is the back veneer. (These stamps do not appear on cabinet-grade *hardwood* panels, which are more expensive and would never be used for a project that's to be painted.)

The higher the combination of grades in a softwood stamp, the more you'll pay for the panel. So it's important to know which quality you need for each project. If you're using sanded plywood outdoors where both sides will show, choose A-B Exterior. If the appearance of one side is less important, use A-C. For storage bins in the carport, B-B or B-C is good enough.

VENEER GRADES

N	Smooth surface "natural finish" veneer. Select, all heartwood or all sapwood. Free of open defects. Allows not more than 6 repairs, wood only, per 4 × 8 panel, made parallel to grain and well matched for grain and color.
A	Smooth, paintable. Not more than 18 neatly made repairs, boat, sled, or router type, and parallel to grain, permitted. May be used for natural finish in less demanding applications.
B	Solid surface veneer. Shims, circular repair plugs and tight knots to 1 inch across grain permitted. Some minor splits permitted.
C **Plugged**	Improved C veneer with splits limited to ⅛ inch width and knotholes and borer holes limited to ¼ × ½ inch. Admits some broken grain. Synthetic repairs permitted.
C	Tight knots to 1½ inch. Knotholes to 1 inch across grain and some to 1½ inch if total width of knots and knotholes is within specified limits. Synthetic or wood repairs. Discoloration and sanding defects that do not impair strength permitted. Limited splits allowed. Stitching permitted.
D	Knots and knotholes to 2½ inch width across grain and ½ inch larger within specified limits. Limited splits allowed. Stitching permitted. Limited to Interior, Exposure 1 and Exposure 2 panels.

If you're planning to paint your project (as is the case with most of the prize projects shown in this book), you'll want MDO (Medium Density Overlaid). A resin-treated fiber surface is heat-fused to the panel face, so MDO won't show those checks that often appear on other plywood panels after weather exposure. Though it's an exterior-type panel, MDO makes equal sense for painted projects indoors—especially those that will get rough use, such as children's furniture.

Other outdoor panels that can be at home indoors are the 303 plywood sidings, a sampling of which is shown on page 5. Though intended for the weather envelope of a house, these panels can lend a novel touch to smaller constructions, indoors and out, since they offer a variety of special textures. But they are "good-one-side-only" panels, and not all the patterns and textures shown are available nationally.

For an indoor project where both sides will show, use A-B, if it's available at your yard. Very few yards stock top-quality A-A softwood ply, but a B-rated face is only slightly inferior. When only one side will show, as on a tabletop, economize with A-D Interior or A-C Exterior (A-C is not available as an interior type).

If you want to master the many varieties of plywood available, start with a study of the grade charts on pages 6–7. They cover all the grade stamps you're likely to find at a lumberyard, and were supplied to us by the American Plywood Association.

What thickness should you choose? You'll find ¼", ⅜", ½", ⅝", and ¾" are the most common, though several grades come in thicknesses up to 1⅛". If your shelf span will be 24" or less, ½" plywood is thick enough for most storage. But for heavy power tools or longer spans, go to ⅝" or ¾". When using plywood for shelving, run the face grain across the supports.

PLYWOOD TEXTURES

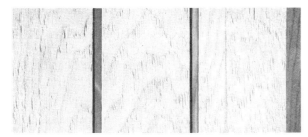

TEXTURE 1-11: Panel has shiplapped edges and ¼″-deep grooves ⅜″ wide, either 4″ or 8″ OC. Name refers to parallel grooving. Most T 1-11 is ⅝″ thick. Also available with scratch-sanded, rough-sawn, overlaid, and other surfaces, in Douglas fir, redwood, cedar, pine, lauan, and other species.

KERFED ROUGH-SAWN: Rough-sawn surface, as above, with narrow grooves added for a planked effect. Long edges shiplapped for continuous pattern. Grooves typically 4″ OC, but also available in spacings in multiples of 2″. Generally comes in ¹¹⁄₃₂″, ⅜″, ½″, ¹⁹⁄₃₂″, ⅝″, and ¾″ thicknesses.

CHANNEL GROOVE: Shallow grooves (typically ¹⁄₁₆″ deep and ⅜″ wide) cut into faces of ⅜″-thick panels, 4″ or 8″ OC (other groove spacings on special order, as are ¹¹⁄₃₂″ and ½″ thicknesses). Available in same surfaces and species as T 1-11, but is thinner panel for where ⅝″ isn't needed.

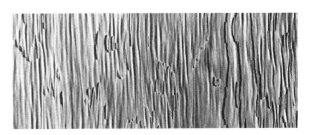

BRUSHED: Relief-grain surfaces accent the natural grain pattern by brushing out softer wood. Available in Douglas fir (shown), redwood, cedar, and other species, generally in ¹¹⁄₃₂″, ⅜″, ½″, ¹⁹⁄₃₂″, ⅝″, and ¾″ thicknesses. Texture lends itself to striking two-tone finishes.

REVERSE BOARD-AND-BATTEN: Deep, wide grooves cut into brushed, rough-sawn, coarse-sanded, or other textured surfaces to suggest batten effect with deep shadow lines. Grooves are ¼″ deep, 1″ to 1½″ wide, spaced 8″, 12″, or 16″ OC. Panels come ¹⁹⁄₃₂″, ⅝″, and ¾″ thick, in redwood, cedar, fir, pine, lauan.

STRIATED: Fine, random-width, closely spaced grooves run full length of panel—ideal for concealing nail heads and joints; hides checking, as well. Lends itself to novel two-tone finishes. Also available; corrugated plywood, with broad uniform grooving for a distinctive shadow line.

FOR EXTERIOR USES

GRADE (EXTERIOR)	FACE	BACK	INNER PLIES	USES
A-A	A	A	C	Outdoor, where appearance of both sides is important.
A-B	A	B	C	Alternate for A-A, where appearance of one side is less important. Face is finish grade.
A-C	A	C	C	Soffits, fences, base for coatings.
B-C	B	C	C	For utility uses such as farm buildings, some kinds of fences, etc., base for coatings.
303® Siding	C (or better)	C	C	Panels with variety of surface texture and grooving patterns. For siding, fences, paneling, screens, etc.
T 1-11®	C	C	C	Special 303 panel with grooves ¼" deep, ⅜" wide. Available unsanded, textured, or MDO surface.
C-C (Plugged)	C Plugged	C	C	Excellent base for tile and linoleum, backing for wall coverings, high-performance coatings.
C-C	C	C	C	Unsanded, for backing and rough construction exposed to weather.
B-B Plyform	B	B	C	Concrete forms. Re-use until wood literally wears out.
MDO	B	B or C	C	Medium Density Overlay. Ideal base for paint; for siding, built-ins, signs, displays.
HDO	A or B	A or B	C-Plugged or C	High Density Overlay. Hard surface; no paint needed. For concrete forms, cabinets, counter tops, tanks.

FOR INTERIOR USES

GRADE (INTERIOR)	FACE	BACK	INNER PLIES	USES
A-A	A	A	D	Cabinet doors, built-ins, furniture where both sides will show.
A-B	A	B	D	Alternate of A-A. Face is finish grade, back is solid and smooth.
A-D	A	D	D	Finish grade face for paneling, built-ins, backing.
B-D	B	D	D	Utility grade. For backing, cabinet sides, etc.
C-D	C	D	D	Sheathing and structural uses such as temporary enclosures, subfloor. Unsanded.
UNDERLAYMENT	C-Plugged	D	C and D	For underlayment or combination subfloor- underlayment under tile, carpeting.

GUIDE TO APPEARANCE GRADES OF PLYWOOD[1]

SPECIFIC GRADES AND THICKNESSES MAY BE IN LOCALLY LIMITED SUPPLY.

	Grade Designation [2]	Description and Most Common Uses	Veneer Grade			Most Common Thicknesses (inch)					
			Face	Inner Plies	Back						
Interior Type	N-N, N-A N-B INT-APA	Cabinet quality. For natural finish furniture, cabinet doors, built-ins, etc. Special order items.	N	C	N, A, or B						¾
	N-D-INT-APA	For natural finish paneling. Special order item.	N	D	D	¼					
	A-A INT-APA	For applications with both sides on view, built-ins, cabinets, furniture, partitions. Smooth face; suitable for painting.	A	D	A	¼		⅜	½	⅝	¾
	A-B INT-APA	Use where appearance of one side is less important, but where two solid surfaces are necessary.	A	D	B	¼		⅜	½	⅝	¾
	A-D INT-APA	Use where appearance of only one side is important. Paneling, built-ins, shelving, partitions, flow racks.	A	D	D	¼		⅜	½	⅝	¾
	B-B INT-APA	Utility panel with two solid sides. Permits circular plugs	B	D	B	¼		⅜	½	⅝	¾
	B-D INT-APA	Utility panel with one solid side. Good for backing, sides of built-ins, industry shelving, slip sheets, separator boards, bins.	B	D	D	¼		⅜	½	⅝	¾
	DECORATIVE PANELS— APA	Rough-sawn, brushed, grooved, or striated faces. For paneling, interior accent walls, built-ins, counter facing, displays, exhibits.	C or btr.	D	D		⁵⁄₆	⅜	½	⅝	
	PLYRON INT-APA	Hardboard face on both sides. For counter tops, shelving, cabinet doors, flooring. Faces tempered, untempered, smooth, or screened.		C & D					½	⅝	¾
Exterior Type	A-A EXT-APA	Use where appearance of both sides is important. Fences, built-ins, signs, boats, cabinets, commercial refrigerators, shipping containers, tote boxes, tanks, ducts. (4)	A	C	A	¼		⅜	½	⅝	¾

(1) Sanded both sides except where decorative or other surfaces specified.
(2) Can be manufactured in Group 1, 2, 3, 4, or 5.
(3) C or better for 5 plies. C Plugged or better for 3 plies.
(4) Can also be manufactured in Structural 1 (all plies limited to Group 1 species) and Structural 11 (all plies limited to Group 1, 2, or 3 species).
(5) Also available as a 303 Siding.

Grade Designation (2)	Description and Most Common Uses	Veneer Grade			Most Common Thicknesses (inch)					
		Face	Inner Plies	Back						
A-B **EXT-APA**	Use where the appearance of one side is less important. (4)	A	C	B	¼		⅜	½	⅝	¾
A-C **EXT-APA**	Use where the appearance of only one side is important. Soffits, fences, structural uses, box-car and truck lining, farm buildings. Tanks, trays, commercial refrigerators. (4)	A	C	C	¼		⅜	½	⅝	¾
B-B **EXT-APA**	Utility panel with solid saces.	B	C	B	¼		⅜	½	⅝	¾
B-C **EXT-APA**	Utility panel for farm service and work buildings, boxcar and truck lining, containers, tanks, agricultural equipment. Also as base for exterior coatings for walls, roofs. (4)	B	C	C	¼		⅜	½	⅝	¾
HDO **EXT-APA**	High Density Overlay plywood. Has a hard, semi-opaque resin-fiber overlay both faces. Abrasion resistant. For concrete forms, cabinets, counter tops, signs, tanks. (4)	A or B	C or C plgd	A or B			⅜	½	⅝	¾
MDO **EXT-APA**	Medium Density Overlay with smooth, opaque, resin-fiber overlay one or both panel faces. Highly recommended for siding and other outdoor applications, built-ins, signs, displays. Ideal base for paint. (4)(5)	B	C	B or C			⅜	½	⅝	¾
303 **SIDING** **EXT-APA**	Proprietary plywood products for exterior siding, fencing, etc. Special surface treatment such as V-groove, channel groove, striated, brushed rough-sawn and texture-embossed MDO. Stud spacing (Span Index) and face grade classification indicated on grade stamp.	(3)	C	C			⅜	½	⅝	
T 1-11 **EXT-APA**	Special 303 panel having groves ¼″ deep, ⅜″ wide, spaced 4″ or 8″ o.c. Other spacing optional. Edges shiplapped. Available unsanded, textured and MDO.	C or btr.	C	C				19/32	⅝	
PLYRON **EXT-APA**	Hardboard faces both sides, tempered, smooth or screened.		C					½	⅝	¾
MARINE **EXT-APA**	Ideal for boat hulls. Made only with Douglas fir or western larch. Special solid jointed core construction. Subject to special limitations on core gaps and number of face repairs. Also available with HDO or MDO faces.	A or B	B	A or B	¼		⅜	½	⅝	¾

Exterior Type

HOW TO WORK WITH PLYWOOD

Plywood is so popular for shop projects because it's easy to work—if you work it *right*. For many projects—including those in this book—it's much faster to use than solid lumber.

In this chapter, we demonstrate the basic techniques of cutting, sanding, planing and joining plywood pieces—plus such specific cabinetmaking skills as making plywood drawers and installing plywood shelves. Although many standard woodworking practices apply to plywood, (you use many of the same tools, adhesives, fasteners and finishes as with lumber) plywood remains a distinct material that requires some special attention.

One example: Because it's a sandwich of thin veneers, plywood is more subject to splintering than solid lumber, and edge treatments become a major concern. Exposed plies have a very different appearance from the surface veneers, and the front and back of the panel may differ greatly. (When you're working an ordinary board, on the other hand, it's much the same all the way through.)

So: When drilling through a plywood panel—even with twist drills, but especially with spade bits—it's best to support the area firmly with back-up scrap you can drill into. Even better, drill from one side of the panel until the point of the bit pierces the other face, then complete the hole from that side.

When laying out panels, following the diagrams we give for each project in this book, draw directly on the plywood (which face you mark depends on the saw you're planning to use, as shown on the following pages; for hand-sawing, mark the best face and keep it up as you cut with a 10-to-15-point crosscut saw). Use a sharp pencil, a carpenter's square and a good straightedge. For circles or rounded corners, use a compass—never freehand a curve. And be sure to check the width of your saw cut and allow for this "kerf" when plotting dimensions.

You might want to take this book to your lumber yard or home center so you can show the staff your project layout; they may offer to do a basic cut or two for you, to make the panel easier for you to cart home. Even if they don't, you'll want your first cuts to be the basic ones that reduce the big 4 × 8-ft. (or 4 × 4-ft.) panel to easily-handled pieces.

Most plywood panels are sanded smooth in manufacture— that's one of their time-saving advantages. As with lumber, *never* sand plywood faces across the grain. If your factory-sanded surface needs further smoothing, first apply a sealer to raise the grain and to protect the softwood streaks. Edges are another matter: Cut edges will always need sanding. Use 1-0 or finer paper *before* you apply any sealer or undercoat. Where the edge is to be glued to another piece, leave it a bit rough to give some "tooth" for the glue.

Assembling a plywood project isn't much different from assembling one made of lumber. For indoor use any woodworker's white glue is fine—but don't use it for joints that will be exposed to the weather; instead use a waterproof glue such as resorcinol or exterior epoxy.

It's usually logical to build by section. Drawers, cabinet cases, compartments, leg units—all should be handled as separate assemblies. Always check for a good fit by holding pieces together *before* applying glue—contact should be made uniformly at all points for best strength. Mark your nail locations along the edge of any piece to be nailed; where the nails are close to the edge, predrill using a bit slightly smaller than the nail. *Always* predrill for screws, boring a *clearance* hole for the unthreaded portion of the screw and a *pilot* hole (of a diameter less than the screw) for the threaded portion. If you don't want the heads to show, use flathead screws and countersink the clearance hole before you drive them. Sink the head below the surface, fill with wood dough and sand smooth.

In finishing, little surface preparation is necessary. If an opaque finish (paint, enamel) is to be used, cover knots, pitch streaks and patches with shellac or other sealer. Brush on an oil-based primer; there are special plywood primers that are pigmented to help hide the "wild grain." Gloss and semi-gloss enamel topcoats provide a washable surface; it can be oil-based, alkyd or latex, providing it is compatible with the primer.

Applying a "natural" (transparent) finish to softwood plywood is trickier. Select your panels carefully for a regular grain pattern and a minimum of face-patches. Brush on varnish (gloss or satin), following label directions. You can subdue grain or patches with a light stain. Dark stains, such as walnut, rarely look good on fir plywood—that zig-zag grain pattern simply doesn't *look* natural when heightened with stain.

For exterior projects, use a top-quality paint or stain to protect against weathering. End grain absorbs and loses moisture rapidly, so seal all edges before exposure. You can use any paint primer to seal edges of projects to be painted, or a water-repellent preservative for panels that will be stained. Brushing is the best method for thorough coverage outdoors.

CUTTING PLYWOOD

GOOD FACE UP is the rule to minimize splintering of the best veneer when you are using a saw that cuts on the *down* thrust (such as this table saw or a radial-arm saw—or any standard handsaw). Whenever the teeth cut on their way *down* they'll chisel a clean top edge but may splinter the bottom face—the "C" face, for example, on an A-C panel.

GOOD FACE DOWN is the rule when you're using a saw that cuts on the *up* thrust, such as the portable circular saw shown—or an electric saber saw. As you feed the teeth of such blades into the work, they slice first into the lower face of the panel and may tend to break out the top veneer. For such sawing, lay out your cut lines on rear of panel.

This was the result when the lumberyard cut a 4 × 8 panel in half for easy transport. The yard's panel saw was coarse-toothed and dull, like blade shown, and the yardman was untrained enough to saw with the good face *up*, so about 2½ inches of face veneer on both sides of the cut became instant waste.

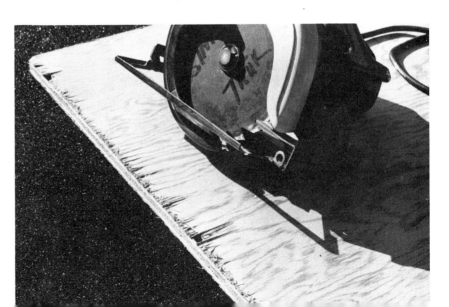

SANDING AND PLANING

For smoothest finish on softwood ply, sand lightly *after* you've brushed on a sealer (as here) or a prime coat. Sanding of uncoated panels should be confined to the edges since most "appearance-grade" plywood is smoothed in manufacture and further surface sanding will remove softer wood in the grain pattern. A sanding block will prevent gouging.

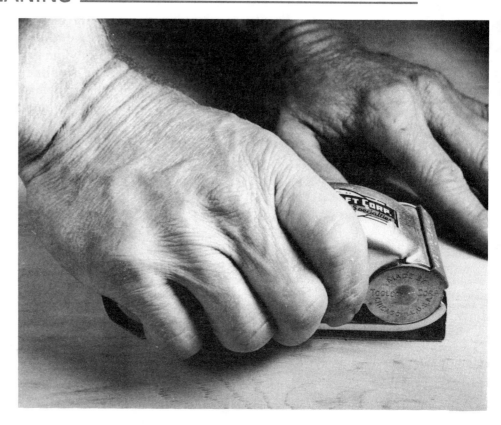

Planing plywood edges won't usually be necessary, if you make your cuts accurately and with a sharp saw blade. But if planing is required, work from both ends of the edge toward the center, to avoid tearing out plies at the end of the cut. Use a plane with a sharp blade set for a very shallow cut and work in short strokes.

PATCHING EXPOSED EDGES

Gaps are common in core plies since lower-quality veneers with knot-holes and splits are used. These voids are exposed when cuts are made, as in photo above.

If edge will be finished natural, best patch is with a scrap of matching wood, coated with glue and held in place with a strip of masking tape until glue sets.

After tape is removed, sand the edge smooth with fine paper wrapped around a sanding block. It's best, of course, to fill other voids in exposed edges at same time.

To prevent any further damage to the edge during assembly of the project, apply a sealer—or finish-coat edges with a quality varnish, as here. Don't coat edges to be glued.

CONCEALING PLYWOOD EDGES

Adhesive-backed, peel-and-stick wood veneer is the easiest way to hide plies. This type bonds when a warm iron is pressed against the tape. Smooth edges give the best bond.

For a well-anchored edging (ideal for tabletops) buy or make T-moldings of a matching softwood (or contrasting hardwood) and drive the glued flange into the kerfed edge.

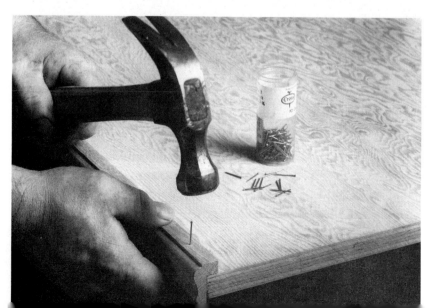

To bulk up an edge, as for a countertop, buy edge molding and fasten it with glue and brads. This is a good way to conceal the edge of applied laminate, but omit the brads.

For a carved effect, buy embossed molding strips from a woodcraft supply house and stain to match or contrast with the surface finish. These strips can be glued in place.

Metal or plastic channel makes a durable edging, if you can find a size that fits the panel thickness snugly. Just drive it in place, using a block to prevent hammer mars.

When you plan to apply plastic laminate to a panel, one method is to cement a matching strip on the front edge first, then brush contact cement on the surface and lap top sheet.

PROTECTING CORNERS

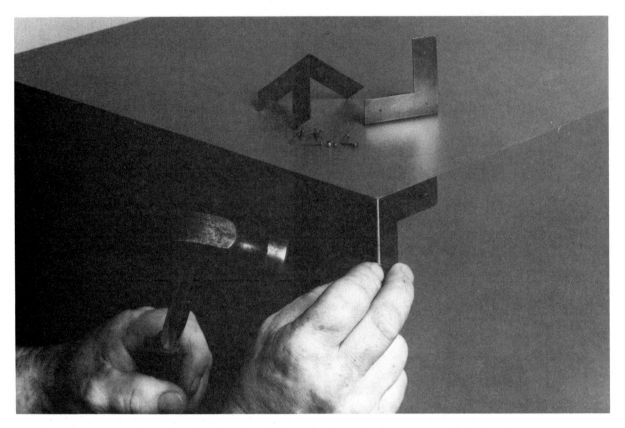

These mating edges have been mitered to hide all plies. This leaves a rather fragile corner, so double-L's of metal are tacked in place for a decorative touch.

Even more familiar are brass-plated footlocker corners like these. Such decorative hardware can be purchased from woodworker's supply catalogs or well-stocked hardware stores.

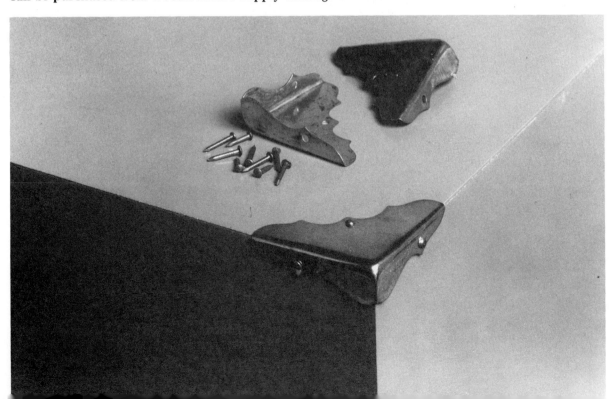

ASSEMBLY JOINTS

Butt joints like the one at left are simplest to make and can be sturdy if panels are ¾″ thick. For thinner panels, use a reinforcing block or nailing strip like the one at the right. These are glued and nailed into the corner, and avoid any nailing into panel edges. All butt joints will be stronger if glued rather than just nailed or screwed.

Frame construction, as with these 2 x 2s, lets you reduce overall project weight and cost by using thinner plywood. Glue-nailing a frame to a plywood panel greatly increases its strength. Note that a recess has been left along the top edges to take a ¾″ thick "lid", which can be nailed or screwed directly to the lumber frame.

In frameless assembly, where nails are to be driven into panel edges—and thus must be placed close to the edge of the top panel—it's best to pre-drill, using a bit slightly smaller in diameter than the nail to be used. Finishing nails are best since heads can be set below the surface and covered with wood dough before the finish is applied.

Rabbet joints are neat and strong—and easy to cut on a table saw or with a router. Note here that the exposed edge of the top panel is greatly reduced by the rabbet, and could even be left natural as a design element. The shoulder of the rabbet "locates" the side panel for glue-nailing. As before, pre-drilling is recommended to avoid splitting.

FASTENERS FOR PLYWOOD

Finishing nails are best for most projects in this book—size determined by panel thickness. Used with glue, all the nails shown here will produce strong joints. From left: 6d casing or 8d finish nails for ¾″ ply; 6d or 8d finish for ⅝″; 4d or 6d for ½″; 3d or 4d for ⅜″ and, for the ¼″ ply at the top, use ¾″ or 1″ brads or 3d finish. For attaching backs, where appearance isn't important, use 1″ blue lath nails (top row, center). Whenever you want a heavier nail for a job, substitute a casing name for a finishing nail of the same length.

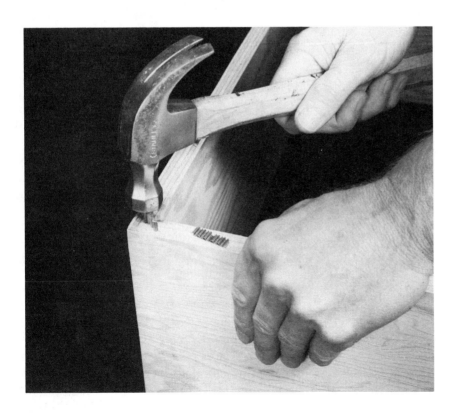

Corrugated fasteners can reinforce miter joints like this one joining two ¾″ panels. They draw the joint tight and hold it while your glue sets. Other fasteners you'll find useful for certain plywood projects include bolts with nuts and washers—particularly for installing surface hinges where short screws could pull loose in heavy use.

Flathead wood screws are best where nails won't provide adequate holding power. Use them with glue unless you plan to take the project apart later. Screws are shown on the panel thickness they're best for, but length is important, too; choose screws from the chart on page 19. Sizes shown are minimums—use longer screws when the work allows. Always pre-drill a clearance hole for the shank and a lead hole for the threaded portion. Countersink the clearance hole so the head sets below the surface and can be capped with a wood-plug or dough.

PLYWOOD THICKNESS	SCREW LENGTHS	SCREW SIZE	DRILL SIZE FOR SHANK	DRILL SIZE FOR ROOT OF THREAD*
3/4"	1½"	#8	11/64"	1/8"
5/8"	1¼"			
1/2"	1¼"	#6	9/64"	3/32"
3/8"	1			
1/4"	1	#4	7/64"	1/16"

*If splitting is a problem (as in edges) make hole for threaded portion 1/64" larger (9/64", 7/64", 5/64" respectively).

ATTACHING BACK PANELS

Set-in backs are common on cabinets and other storage units, but this calls for rabbeting all the sides—only practical with power tools. Unit at left has rabbet just deep enough to take thickness of back panel. For larger assemblies which may not be perfectly plumb, a deeper rabbet is best. One on right is ½" deep so ¼"-thick panel leaves lip.

If you're building only with hand tools, create a rabbet by gluing and bradding on quarter-round molding as shown. This unit, with mitered corners, needed a flush back panel; molding strips (also mitered, for neatness) were placed so their flats came ¼″ below the edges of the side pieces. The ¼″-thick back is being set into place for bradding.

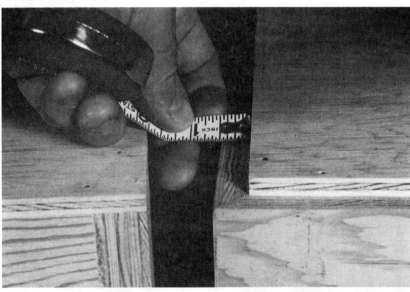

Here are two methods for applying backs without rabbets or moldings. On the left, the back is nailed on the sides, with all edges flush—fine when the project will be painted or veneered. At right, the back has been recessed ½″ from all edges and won't be seen when the cabinet is placed against a wall or set on the floor.

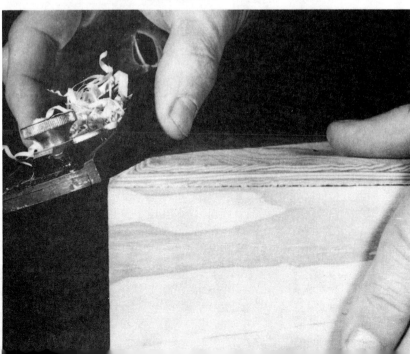

Finally, if you prefer a full-size back panel glued and nailed to the edges of the sides, but don't want those plies to show, simply bevel all the edges to make them inconspicuous. Here, a ⅜″-thick panel has been installed and is having its edges beveled with light strokes of a block plane. No power tools are needed for this technique.

INSTALLING SHELVES

Dado joints, quickly made with a power saw, produce a neat shelf assembly. Use a dado head with blades shimmed out to produce a face groove slightly wider than the shelf thickness. Here the groove is 5/16″ deep across a 3/4″ side, to take a 1/4″ shelf. If shelf must be removed later, let friction hold it in place. For permanent assembly, glue it.

Adjustable shelves are easily placed—and the only power tool you need is a portable drill to bore the blind sockets for the plugs on the L-shaped brackets. You need four for each shelf, and extra holes above and below let you raise or lower them. You can also screw on four slotted shelf standards into which supports may be plugged at any height.

MAKING ALL-PLYWOOD DRAWERS

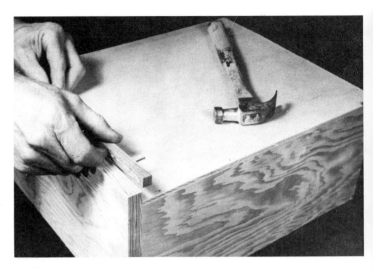

Only hand tools to work with? This drawer (shown upside down) was made with a saw and hammer. Butt joints are glued and nailed and a finger notch has been cut in the top edge of the front to avoid the need for a pull. The bottom edge extends down to cover the front edge of the bottom panel, which is ½" thick for rigidity (for smaller drawers: ⅜").

You can get by with ¼" bottoms for larger drawers if you make the drawer front long enough to cover a reinforcing strip, glued and nailed as shown—but you'll have to provide clearance for this strip to let the drawer close flush, and the drawer will have to ride on hardware slides or one of the groove methods shown in the next photos.

Power tools make it even easier to build sturdy drawers. This one has all the rabbets, grooves and dadoes any cabinetmaker could want. The side being placed (with the rabbeted drawer front to the right, here) is grooved on the outer face for a slide. On the inner face, it's dadoed to receive the back panel and has a groove (unseen) to take the bottom.

Here are two types of guides that call for power tools. The first drawer side has been grooved before assembly to fit over a strip glued to the side of the cabinet. The side of the other drawer has the strip glued to *it*, while a mating groove has been plowed across the inner face of the cabinet. These drawers will slide best if the guides are waxed.

PROJECTING BOTTOM GUIDES

Another hand-tool assembly, this drawer has a projecting bottom that doubles as a slide. Use a ⅜″ or ¼″ panel and let it project ⅜″ beyond both sides of the drawer. A front panel of the same width is being glue-nailed here to the front edges of the sides; it will cover the front edge of the bottom. Sand the projecting edges and apply paraffin.

After the carcass is assembled, slide the projecting edges of the drawers into grooves formed by gluing pieces of ⅜″ plywood to the inner faces of each side. Leave a gap just wide enough to take the lip without binding, and set these panels ¾″ in from the front edge of the sides. That lets the drawer front set flush within these side edges.

Power tools let you make a simpler, lighter version of this same drawer. The ¾″ front has been rabbeted on three edges—1½″ wide across the two ends to create a shoulder against which the side pieces rest. The ¼″ bottom panel is wide enough to project ⅜″ on each side when it is put in place, and long enough to cover the edge of the back.

This drawer slides in ⅜″-deep grooves that were dadoed across the inner faces of the cabinet sides before assembly. The drawer front extends an additional ⅜″ beyond the bottom at each side to cover the edge of the cabinet sides when closed. Note the groove above the side panel to take the bottom of the next-higher drawer in this chest.

INDOOR PROJECTS

SAWBUCK TABLE

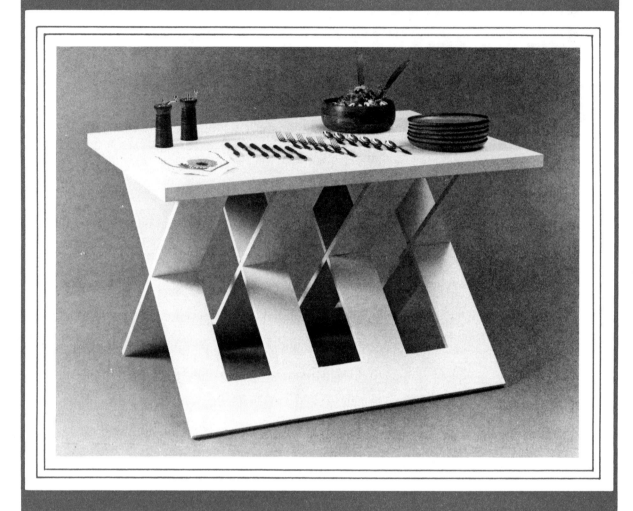

Plywood panels are especially appropriate for large tables. This sawbuck with a 36″ × 48″ top could scarcely have been built from anything else. It was our earliest top prize winner, from our first contest, and all parts come from a single ¾″ sheet, as shown in the cutting diagram on the next page. You need no additional material except glue and paint.

Each leg of the sawbuck support panels is slotted half-way through to slip into the notch in its mating leg. The top end of each leg must be beveled 45 degrees so it will seat flat against the tabletop. This sawbuck assembly is held firmly in place by means of mitered apron pieces that form a frame around the tabletop's underside, and give a more substantial look to the panel as a bonus. Note that the inner edges of the two longer apron

Norbert G. Marklin
St. Louis, MO

pieces must be beveled to 45 degrees—before assembly—so as to create a "pocket" for the beveled legs to fit into.

On a level floor, the table will be sturdy enough for any normal load—as a showcase for houseplants, say, or magazines. It is ideal as a buffet-server for an informal meal, or can be set up as a party bar. The beauty of it is that once it's served its purpose, you can easily disassemble it by simply pinching the leg panels together

PANEL LAYOUT

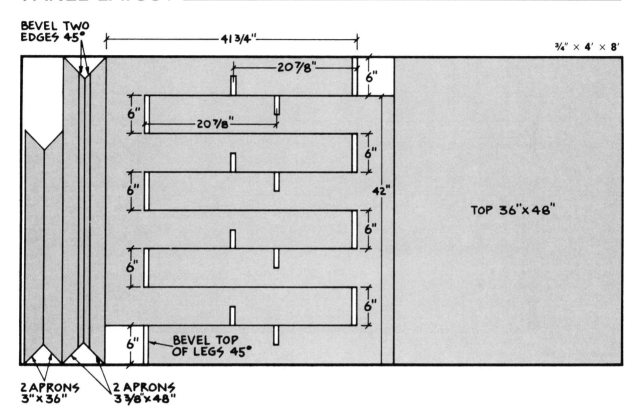

BEVEL TWO EDGES 45°

41 3/4"

3/4" × 4' × 8'

20 7/8"

6"

6"

20 7/8"

6"

6"

6"

42"

TOP 36"x 48"

6"

6"

6"

BEVEL TOP OF LEGS 45°

2 APRONS 3" x 36"

2 APRONS 3 3/8"x 48"

⬛ MATERIALS LIST ⬛

Quantity	Description
1 4 x 8 panel	¾" plywood, Medium Density Overlaid (MDO), A-B or A-D Interior, or A-B or A-C Exterior
As required	3d finishing nails for glue-nailing apron to table top
As required	White glue (urea resin type recommended) for glue-nailing apron
As required	Wood dough or other filler for filling any voids in cut edges, fine sandpaper for smoothing cut edges and filler, and primer and enamel for finishing

SIDE VIEW

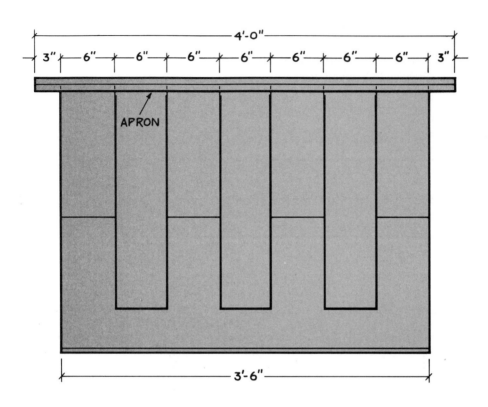

APRON

4'-0"

3" 6" 6" 6" 6" 6" 6" 6" 3"

3'-6"

END VIEW

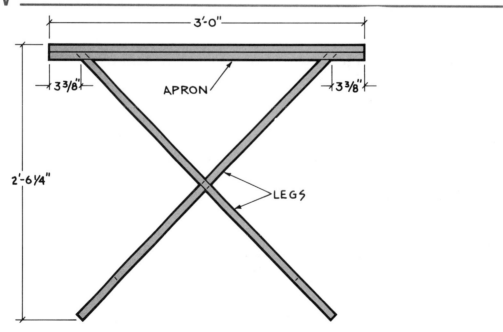

3'-0"

3 3/8"   APRON   3 3/8"

2'-6 1/4"

LEGS

EXPLODED VIEW

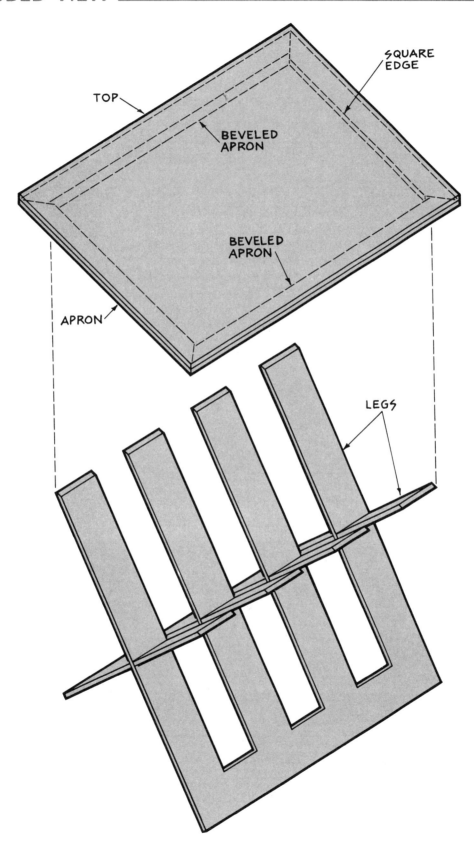

enough so you can lift off the top. Then all you have to store is three flat panels.

We even tried the table for dining; it's fine for two, but seat more than that along its sides and they'll be competing for knee slots. And you can't sit comfortably at the ends.

It's a good idea to attach the apron pieces—especially the long pair, since they must resist shear pressure—with three penny finishing nails spaced about 8 inches apart in staggered rows. But along the inside edges, be sure to keep these back far enough so they don't pierce the bevel. All bevels must be cut (on a tilt-table bandsaw, or with a portable saber saw) after the parts have been straight-cut.

MDO is the ideal panel here if you plan to enamel the parts as shown. Remember that an enamel coat adds to each part's thickness, so make your slots slightly oversize to compensate. The slot-fit is not really critical since the main support comes from the fact the leg assembly is "captured" within that apron frame. This prize-winning table is as simple to construct and as-semble as it is elegant to look at.

END VIEW DETAIL: APRON & LEGS

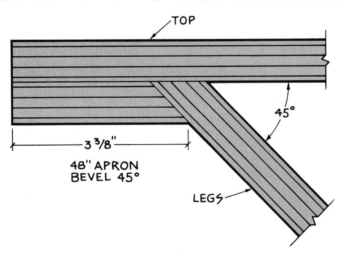

TOP

45°

3 3/8"

48" APRON
BEVEL 45°

LEGS

FRONT VIEW DETAIL: APRON FROM INSIDE

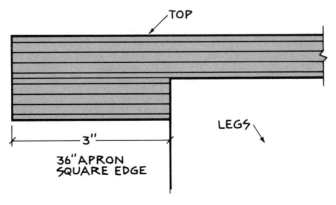

TOP

LEGS

3"

36" APRON
SQUARE EDGE

KNOCKDOWN DINING/SERVING TABLE

Like the previous table, this one has a top with a bulked-up edge—underside framing strips not only increase the thickness, visually, but create a recess into which the leg assembly locks for sturdiness. And, also like the first table, this one slots together for easy store-flat disassembly. This is, again, a particularly elegant design, and one that makes efficient use of its sheet and a half of plywood.

The table is laid out for the simplest possible construction. The designer determined his initial cut sequence to speed the work. The first cut trims off a panel to be set aside for later; the second and third cuts notch out a section that leaves a work-piece consisting of identical halves in mirror image. Your fourth

John Leggitt
Santa Barbara, CA

PANEL LAYOUTS

½″ × **4′** × **8′**

2nd CUT 2nd CUT 1st CUT

72″ 24″

6″ 45″ 15″ 6″ 6″ 6″ 3″ 3″ 3″ 3″

EDGE PIECES 45″x 3″

1″ R. TYP.

15″ TYP.

15″ TYP.

A D C

42″

B

EDGE PIECES 22½″x 3″ EDGE PIECES 22½″x 3″ 3″ TYP.

3″

EDGE PIECES 45″x 3″

A B

6″ EDGE PIECES 22½″x 3″ 3rd CUT EDGE PIECES 22½″x 3″

24″ 6″ 6″ 24″

30″ 30″

½″ × **4′** × **4′**

TABLE TOP 48″x 48″

cut slices this panel on its centerline; you flop the right half under the left, tack them together, and do the balance of the cutting through both halves simultaneously. That way you'll assure duplicate parts for precise assembly. Those 18 3″-wide edge strips (six a full 45″ long, 12 half-length) are glue-nailed in layers to beef up the 4′-square top to a handsome 2″ thickness.

MATERIALS LIST

Quantity	Description
1 4 x 8 panel	½″ plywood, A-B or A-D Interior, A-B or A-C Exterior, or MDO
1 4 x 4 panel	½″ plywood of same grade (Ask your dealer about availability of half-panels.)
Small can	White glue or powdered resin glue for laminating edge strips (waterproof glue if for exterior use)
As required	Wood dough for filling any small voids in plywood cut edges; fine sandpaper for smoothing cut edges and cured wood dough
As required	Finishing materials

SIDE VIEW

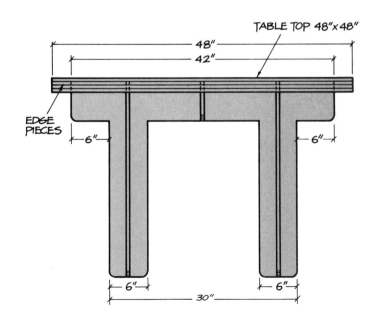

EXPLODED VIEW

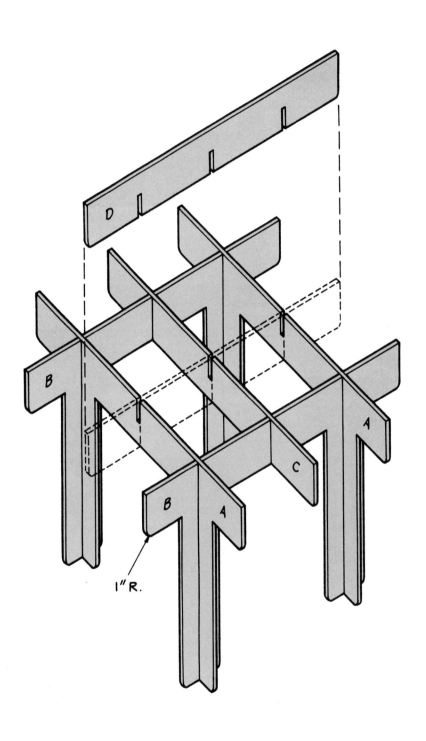

After completing all slots (finish their ends with a sharp ½"-wide chisel), assemble the leg pieces to check for fit. Make any necessary adjustments. Slip *A* pieces down onto *B* pieces to form the four basic legs. Slip *C* into place to form one cross support, add *D* to form the other. When you add the top, its lip locks the leg assembly in position.

Disassemble the table and fill any edge voids with wood dough. When this is dry, sand smooth. Fill and sand the top's laminated edges in the same manner. You're now ready to finish the table. If you use paint, apply a good primer and two coats of a compatible enamel. The designer/builder got a striking effect on the prototype shown by painting the top as three concentric squares. The central 40″ square is a warm grey, framed in a 4″-wide band of off-white; the outer frame (also 4″ wide) is terra-cotta, the color that folds down over the edges and covers the entire leg assembly. For the smoothest enamel coat, of course, you should use MDO plywood. Another finishing possibility would be to paint all surfaces but leave the edges natural, for contrast, coating them with a polyurethane varnish.

The table will comfortably seat four for dining, as shown in the photograph. And the 4′-square top provides a generous work surface for many jobs.

TOP VIEW & EDGE DETAIL

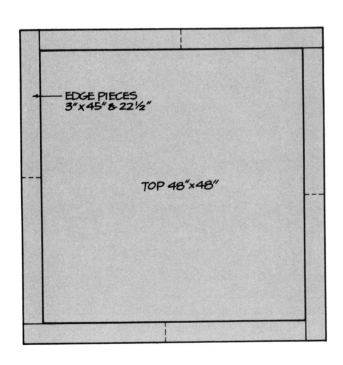

EDGE PIECES
3″ x 45″ & 22½″

TOP 48″x48″

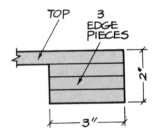

TOP

3
EDGE
PIECES

2′

3″

KNOCKDOWN BUFFET/DINING TABLE

Lack a formal dining table or handy buffet for self-serve parties? Need a quick-to-set-up-and-store-away conference table for committee meetings? Want a family activity table (or one for party games) that can be tucked in a closet when it's not needed?

Here's a good-looking, sturdy table that can seat ten comfortably—and it uses every square inch of a ¾″ thick panel of plywood. (Those four *D* pieces are positioning blocks for the underside of the top.) The table's slot-together design makes it a snap to set up and knock down, yet offers the option of permanent assembly if you prefer to glue all members together.

No workshop skills are required to make this table. If you can use a portable saber saw, you can easily complete the project in

Charles T. Goulding
Elkins Park, PA

one day. Just be sure to start all cuts at the *D* locations after you've laid out all parts on the underside of your plywood panel. To draw the shape of the table top (which also shapes the leg units), tie a string to your pencil. Measure along the string 44¼" and anchor this pivot-point at the center of one long edge of the panel. Scribe an arc as shown, 3¾" from the oposite edge. Repeat this process *from* that opposite edge. Mark points 4½" in from each end, centered 32" apart. With a straightedge, run the ends of your arcs out to these points, then use a compass to scribe a 3" radius at all four corners. Complete the layout as shown and cut out all parts, centering your saw kerf on the lines. Start cuts on any line between parts *D* and *B*, since the curved end of *D* parts gives you smooth access to the cutting lines of the table top. Cut all slots accurately, finishing the ends with a chisel

Lightly sand all cut edges smooth and fill edge voids with wood dough. Sand smooth when dry. Assemble the legs to check the fit of the slots, making any necessary adjustments. Note that slots are not the same depth: The cross braces (parts *C*) extend ¼" above the leg units so that no beveling of the top edges of these angled parts is required.

PANEL LAYOUT

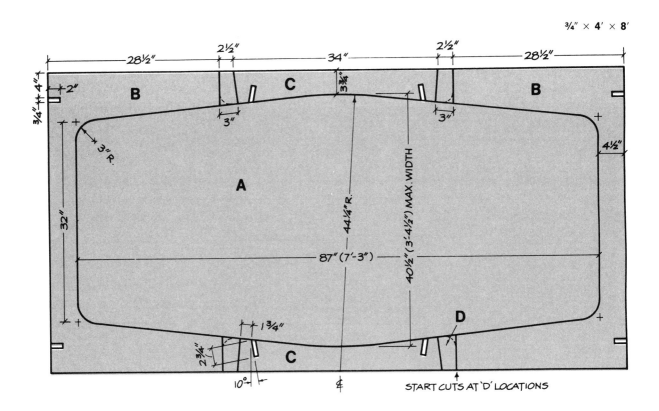

MATERIALS LIST

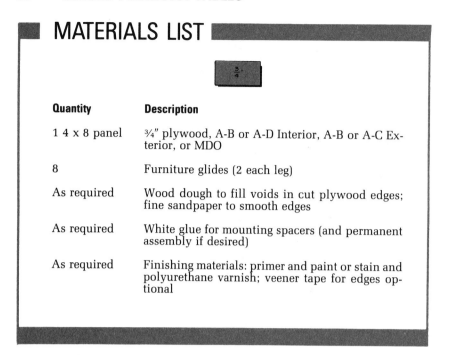

Quantity	Description
1 4 x 8 panel	¾″ plywood, A-B or A-D Interior, A-B or A-C Exterior, or MDO
8	Furniture glides (2 each leg)
As required	Wood dough to fill voids in cut plywood edges; fine sandpaper to smooth edges
As required	White glue for mounting spacers (and permanent assembly if desired)
As required	Finishing materials: primer and paint or stain and polyurethane varnish; veener tape for edges optional

SIDE VIEW

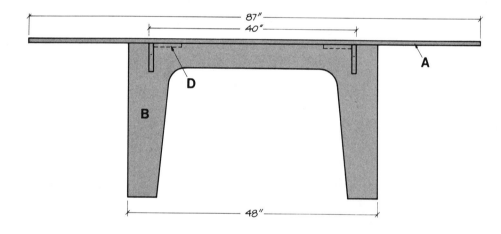

END VIEW

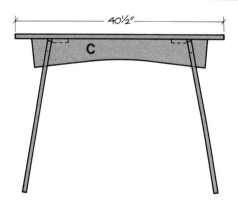

Lay the table top face down, set the assembled legs and cross braces into their centered position and mark locations for the *D* spacers. Glue-nail these spacers in place.

The table in our photo was painted a Mandarin red, but you have many options, including contrasting colors for the top and legs, or leaving all edges in a contrasting color to the surfaces.

The table top has been designed to rest only on the cross braces. If you prefer, you can bevel the top edges of the leg assemblies and lengthen the slots to let the leg units support the table more fully—certainly preferable if you opt for permanent assembly with glued slots.

BOTTOM VIEW

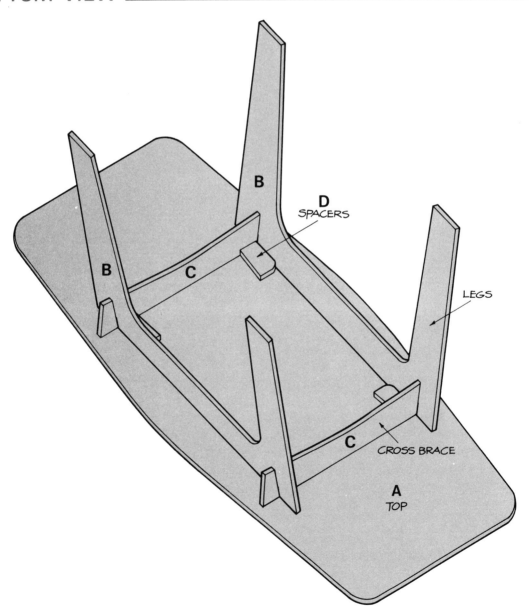

BUFFET/DINING TABLE

Designed for those who need a larger table only for company, this clever unit is 30″ high and 48″ wide. When folded for storage, it's only 4½″ deep. Lift the narrow leaf and you have an 18″-deep buffet—shown on the facing page. Lift the wider leaf and you have a 34½″-deep table that seats two. With both leaves raised, you've got a 4′-square top that seats three or four— or, in a pinch, six. And all from a single sheet of ¾″ plywood, with nothing else to buy except three continuous hinges and the finishing material of your choice.

Lay out the parts as shown and center your saw blade on the layout lines to distribute the kerf equally on all parts. Note that most of the cut-lines shape two parts at once. This whole project can be assembled and painted in one weekend, especially if you

Gregory L. Nelson
Carbondale, IL

40

own an electric screwdriver. Attaching five lengths of continuous hinge can be a bit tedious by hand.

Note that the swing-out supports must clear the barrels of those hinges, so you must provide hardboard shims at their support points. Otherwise, the leaves would drop. Actually, these shims, glued to the underside of the leaves before you apply any finish, help you position the legs quickly, by feel.

After you've cut all members, screw-glue the strips on the underside of the center top, slip the center support between them

and screw-glue it in place as shown in the detail. Now, sand all edges, fill any edge voids with wood dough, and apply your chosen finish before hinging the leaves together. We painted our

MATERIALS LIST

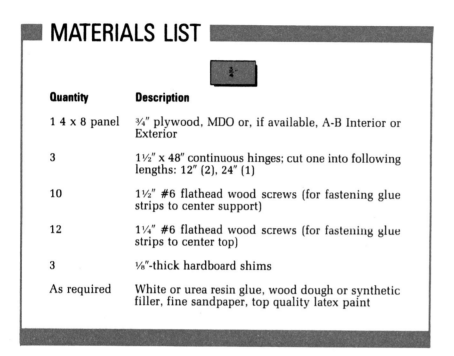

Quantity	Description
1 4 x 8 panel	¾" plywood, MDO or, if available, A-B Interior or Exterior
3	1½" x 48" continuous hinges; cut one into following lengths: 12" (2), 24" (1)
10	1½" #6 flathead wood screws (for fastening glue strips to center support)
12	1¼" #6 flathead wood screws (for fastening glue strips to center top)
3	⅛"-thick hardboard shims
As required	White or urea resin glue, wood dough or synthetic filler, fine sandpaper, top quality latex paint

PANEL LAYOUT

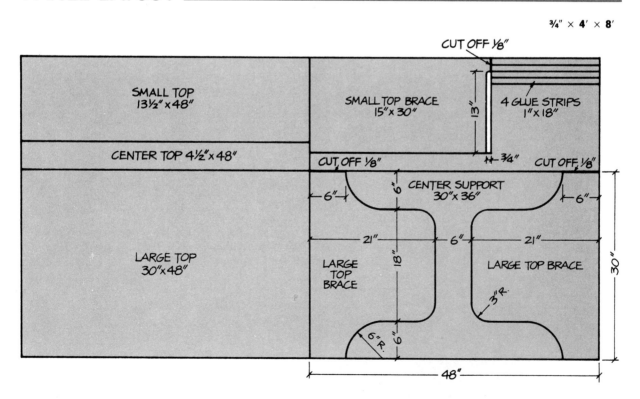

¾" × 4' × 8'

CUT OFF ⅛"

SMALL TOP
13½" x 48"

SMALL TOP BRACE
15" x 30"

3"

4 GLUE STRIPS
1" x 18"

CENTER TOP 4½" x 48"

CUT OFF ⅛"

¾"

CUT OFF ⅛"

6"

CENTER SUPPORT
30" x 36"

6"

6"

LARGE TOP
30" x 48"

21"

6"

21"

18"

LARGE
TOP
BRACE

LARGE TOP BRACE

30"

3" R.

6" R.

6"

48"

prototype a medium blue overall—this isn't a design that lends itself particularly to contrasting edge treatment.

What holds the unit up when it's fully folded? The edge of the deep leaf butts against the floor, just as it does in the unit's buffet mode, stabilizing it. The unit sets up solidly on most carpeting. If the supports are to be swung over a bare floor, you may want to drive small "domes of silence" into their bottom edges.

ISOMETRIC VIEW

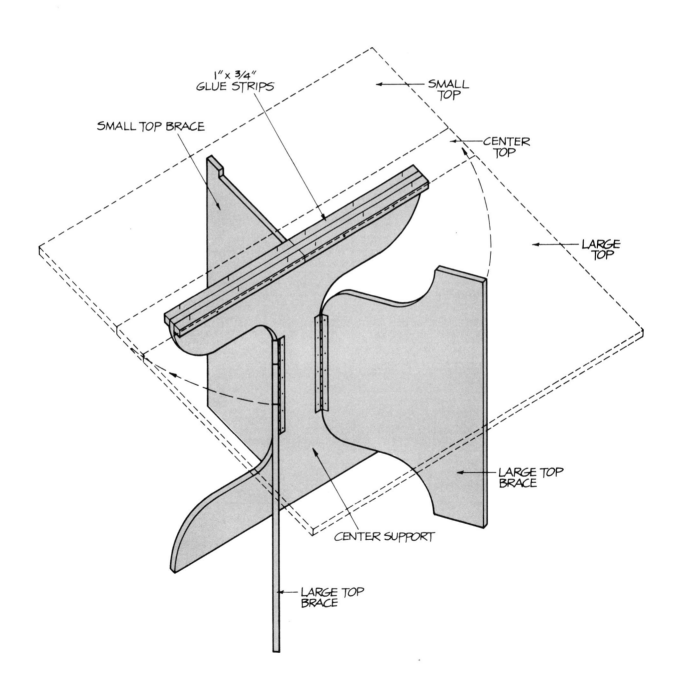

1" x 3/4" GLUE STRIPS

SMALL TOP BRACE

SMALL TOP

CENTER TOP

LARGE TOP

LARGE TOP BRACE

CENTER SUPPORT

LARGE TOP BRACE

BOTTOM VIEW

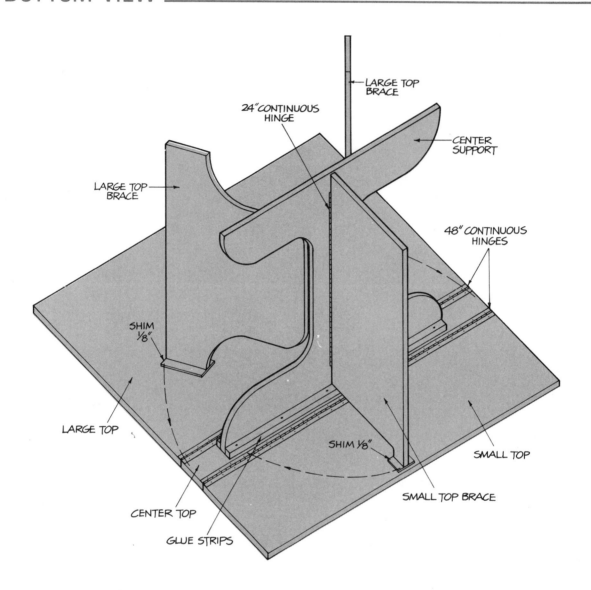

LARGE TOP BRACE

24" CONTINUOUS HINGE

CENTER SUPPORT

LARGE TOP BRACE

48" CONTINUOUS HINGES

SHIM 1/8"

LARGE TOP

SHIM 1/8"

SMALL TOP

CENTER TOP

SMALL TOP BRACE

GLUE STRIPS

HINGE DETAIL

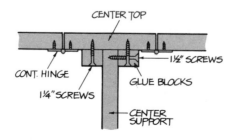

CENTER TOP

CONT. HINGE

1 1/2" SCREWS

1 1/4" SCREWS

GLUE BLOCKS

CENTER SUPPORT

SLOT-TOGETHER COFFEE TABLE

The zero-waste cutting layout for this clever table even provides four positioning blocks for the tabletop's underside. The table's contemporary styling can best be emphasized by painting all edges a lighter color than the surfaces. In our prototype, we painted all surfaces a deep blue and enameled all edges white.

This is another simple project that requires no woodworking skills—just the patient use of a portable saber saw. The leg units slot together for sturdy support of the top, which is held in place by the four small blocks glue-nailed to its underside. The unit disassembles into flat panels for moving or storage.

You'll need a simple school compass set to a 1″ radius to mark most of the corners in the layout. Note how shrewdly the parts

Lori Michelle Bowden
Carbondale, IL

are nested to use every bit of a half sheet of ¾″ ply. After cutting, you can clamp both leg units together and cut all slots in both at once, finishing them to depth with a ¾″ chisel after removing the clamps. Slot the cross members and assemble with the legs; center the assembly on the face-down table top and mark the positions for the four blocks.

PANEL LAYOUT

¾″ × 4′ × 4′

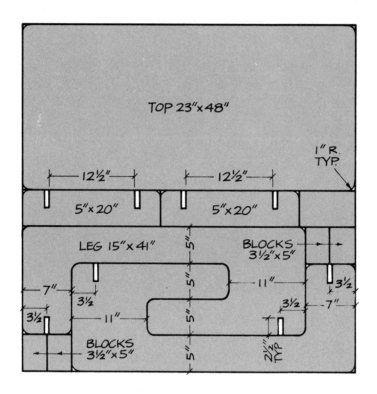

◼ MATERIALS LIST ◼

Quantity	Description
1 4 x 4 panel	¾″ plywood, A-B or A-D Interior, A-B or A-C Exterior, or MDO (Ask your dealer about availability of half panels)
12	3d finishing or 1¼″ flat-head nails for installing blocks
As required	White glue for holding blocks in place and for permanent table assembly, if desired
As required	Fine sandpaper, wood dough for filling any small voids in plywood cut edges, primer and enamel or plastic laminate and edge stripping

BOTTOM VIEW

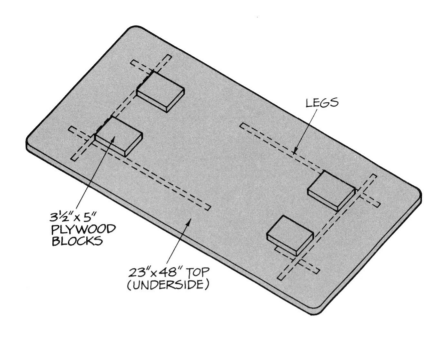

LEGS

3½" x 5"
PLYWOOD
BLOCKS

23" x 48" TOP
(UNDERSIDE)

ISOMETRIC VIEW

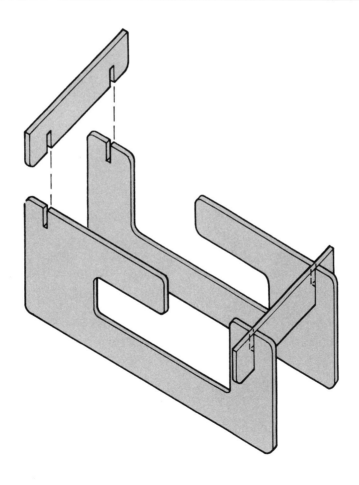

KETTLEDRUM DINING/COCKTAIL TABLE

An exceptionally sturdy combination table results when this kettle-drum base is bolted into recesses in the four crescent feet. Bolt the drum high in these assemblies and you've got a 4'-diameter table for two to four diners (or card players). Pull 12 bolts to drop the drum deeper into the recesses (photo next page) and you've a large, low table for serving beverages.

For easy storage or transport to a new location, you just pop the snap latches that lock the tabletop in place, and disassemble the slot-together drum. All parts are scaled to be nested on one ¾″ panel with virtually no waste.

Lay out parts as shown at right, scribing circles and curves with a bar compass or a straight piece of lath with holes accurately drilled to take a pivot point (a tack or sharp nail) near one end

Kenneth P. Hatfield
Topsfield, MA

and a pencil point spaced at distances of 21″ and 24″ from the pivot. You'll cut neater circles if you use a similar pivot arm to guide your saber saw. To saw the inner circle, drill a starting hole at the slot location. This cut will form a 3″-wide ring that bulks up the edge of the tabletop while forming a recess into which the drum support will fit. Glue and screw the ring to the underside of the top disk. Those footlocker-type latches are optional. With their stationary hook parts screwed to the inner edge of the ring and the snap sections mounted on the four edges of the drum panels, they let you lift the table by its top while you're adjusting the height.

Assemble the four double crescents that form the feet. Note that the bottom tips of each pair are joined by a large dowel glued into drilled recesses (see detail). The dowel ends are slightly angled to bring the unjoined upper tips together; this creates a slight clamping tension when the assemblies are slipped onto the drum panels. Put them in place as shown and drill three holes for the high position, right through each foot-and-drum-panel sandwich.

Now, lower the drum assembly deeper into the recesses until its bottom curves rest on the floor. Be sure the table top is level before drilling three more bolt holes through each crescent, using the previous holes as a template. Choose which sets of height-holes you want to use and insert 12 2½″-long brass bolts. Slip a

PANEL LAYOUT

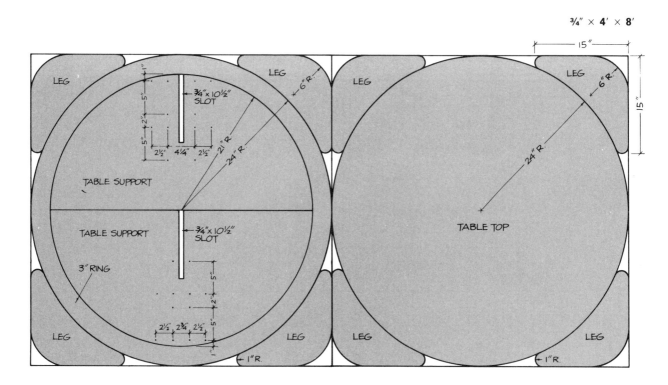

brass washer on each and snug up with a brass wing-nut. You can use slotted round-head bolts or hex heads since tightening is done by finger-turning the wing-nuts. We finished our table with a darker color on the edges of tabletop and feet.

MATERIALS LIST

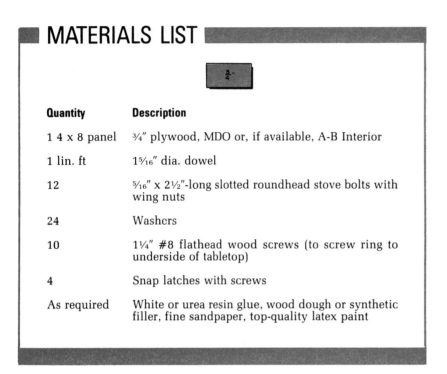

Quantity	Description
1 4 x 8 panel	¾" plywood, MDO or, if available, A-B Interior
1 lin. ft	1⁵⁄₁₆" dia. dowel
12	⁵⁄₁₆" x 2½"-long slotted roundhead stove bolts with wing nuts
24	Washers
10	1¼" #8 flathead wood screws (to screw ring to underside of tabletop)
4	Snap latches with screws
As required	White or urea resin glue, wood dough or synthetic filler, fine sandpaper, top-quality latex paint

SIDE VIEW

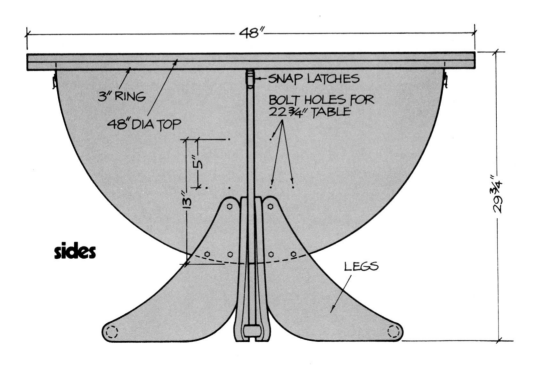

48"

3" RING

SNAP LATCHES

48" DIA TOP

BOLT HOLES FOR 22¾" TABLE

5"

13"

29¾"

sides

LEGS

BOTTOM VIEW

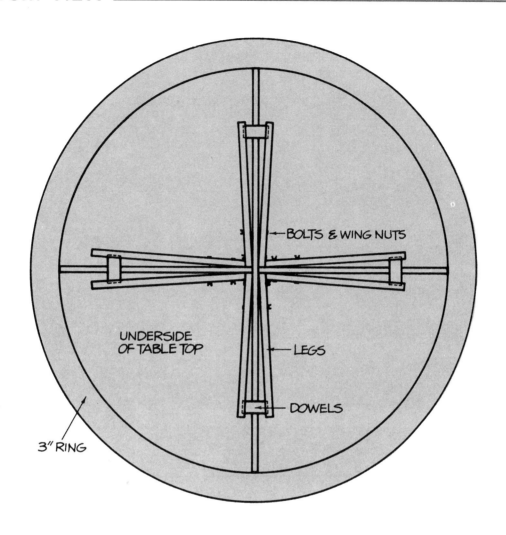

BOLTS & WING NUTS

UNDERSIDE
OF TABLE TOP

LEGS

DOWELS

3" RING

LEG DETAILS

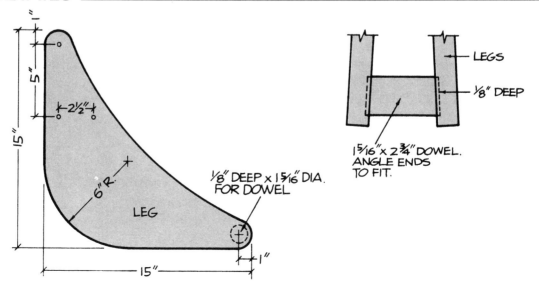

1"

5"

2½"

15"

6" R

LEG

⅛" DEEP × 1⁵⁄₁₆" DIA.
FOR DOWEL

1"

15"

LEGS

⅛" DEEP

1⁵⁄₁₆" × 2¾" DOWEL.
ANGLE ENDS
TO FIT.

PIVOTING DROP-LEAF TABLE

Truly shrewd design lets you get all parts for this clever table out of a half-sheet of ¾″ plywood. A standard lazy Susan bearing (available from hobby shops and woodcraft catalog houses) pivots the central triangle over a tripod base. Positioned as shown above, the three hinged leaves are firmly supported by the projecting legs. But when you pivot the triangle into alignment with the base, the leaves drop neatly *between* the legs.

We two-toned our prototype with contrasting enamels, coating the top and leaves a terra-cotta and running this same color down the edge of each leg. The rest of the base was painted off-white. This has to be one of the handiest and most elegant little tables we've ever built: With the leaves dropped, it takes up only 2½′

Gary Waugh
Truro, NS, Canada

of space and is ideal as a lamp table or plant stand. Turn the top to raise the leaves and you've got a good-sized hexagonal table for serving coffee. It stands about 20″ high.

Both cutting and assembly are simple. The only beveled edges are on the ends of the side panels that form that inner apron. Bevel these ends to 60 degrees and glue and nail each panel to a spacer set at right angles. Glue and toenail these assemblies to the leg members to form the tripod. This creates a solid top platform for mounting the lazy Susan unit: Center it on this platform and drill clearance holes for the mounting bolts. Remove the unit and attach its top plate to the underside of the triangular top—again making certain it's centered. Now turn the entire table assembly upside-down, realign the bottom plate of the bearing with the holes through the platform and anchor it by driving bolts through from the bottom of the spacers, as shown in the diagram. Because of the thickness of the bearing assembly, you'll want to raise the leaf-support points at the outer edge of each leg. The best way is to tap a "dome of silence" or other furniture bumper into each location. You can attach the leaves to the edges of the top triangle with butt hinges as shown (our diagram is a bit of an overkill) or with three continuous hinges.

PANEL LAYOUT

¾″ × **4′** × **4′**

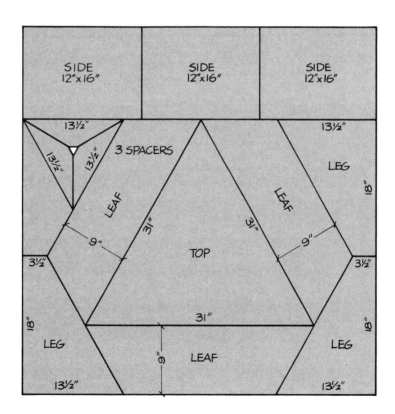

MATERIALS LIST

Quantity	Description
1 4 x 4 panel	¾″ plywood, MDO or A-B Interior
12	2″ hinges with screws
3	Tack bumpers (sized to match height of lazy Susan bearing)
1	6″ lazy Susan bearing
As required	Finishing nails, white or urea resin glue, wood dough or synthetic filler, fine sandpaper, top quality finish

HINGE DETAILS

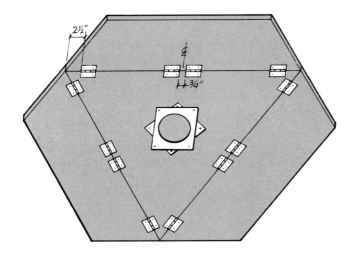

LAZY SUSAN ASSEMBLY

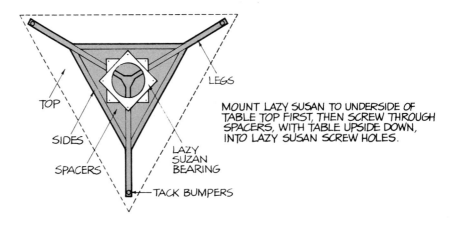

MOUNT LAZY SUSAN TO UNDERSIDE OF TABLE TOP FIRST, THEN SCREW THROUGH SPACERS, WITH TABLE UPSIDE DOWN, INTO LAZY SUSAN SCREW HOLES.

TOP

LEGS

SIDES

SPACERS

LAZY SUZAN BEARING

TACK BUMPERS

ANGLE CUTS FOR SIDES

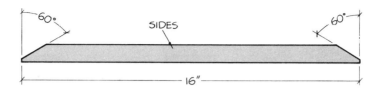

60°   SIDES   60°
16"

EXPLODED VIEW

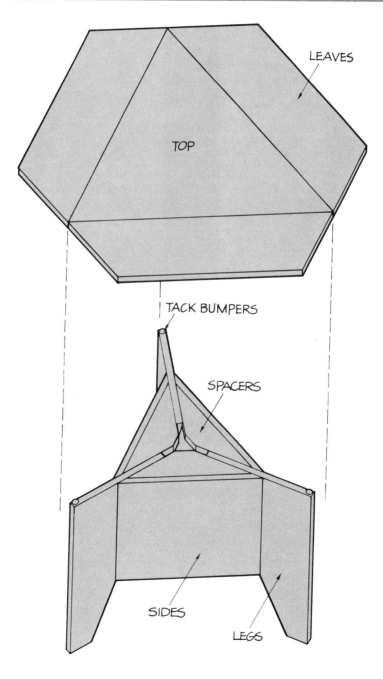

LEAVES

TOP

TACK BUMPERS

SPACERS

SIDES

LEGS

INVERTIBLE COFFEE TABLE

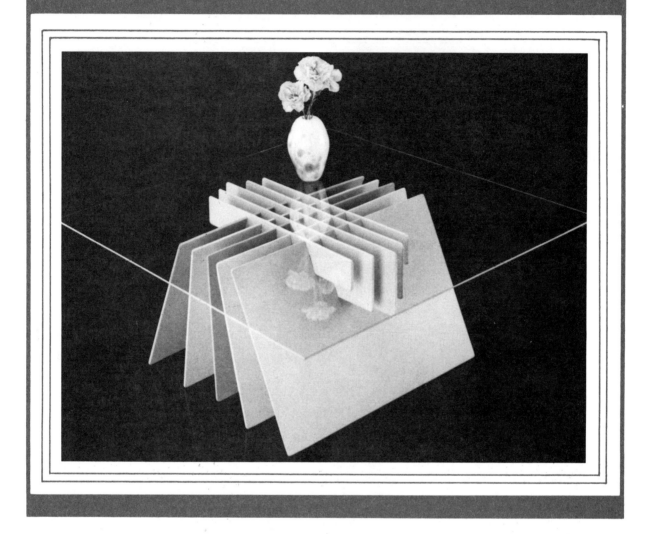

Melissa White
Providence, RI

For sheer elegance, no project in this book tops this one. When you wish to change the design, you just flip the base. Your 3'-square glass or clear plastic top lets you enjoy either pattern: the complex egg-crate shown above or the "opening book" configuration in the smaller photo, next page. The prototype had one face of its "pages" painted in shades of blue-gray and mauve for a truly subtle effect in either position. (The ground-glass edge of the tabletop was also tinted with mauve paint.)

The sophisticated simplicity of this design can be deceptive. Duplicating it should be a challenge to any experienced craftsman. The slots must be precisely angled and all panels accurately cut so that both top and bottom of the assembly will be perfectly flat. You can imagine how many trial-and-error cuts were required to perfect the prototype.

All parts from the base can be cut from a half-panel of ½″ plywood, as diagrammed below. Note that the slots in all five of the large "page" panels are centered on one long edge, but are not cut to identical depths. Once you've cut the plywood sheet up into the ten basic panels, you can tack four of these large panels into pairs, since the two *B* panels and the two *C* panels will be identical; do the slot layout only once and cut them as pairs. Note that the very critical dimensions and angles for the five smaller panels are detailed on the next page. Here, again, the two *E* panels and the two *F* panels may be paired for simultaneous cutting. Make all the slots slightly wider than the ½″ thickness of the plywood, to ease the slot-together assembly—especially if you are planning to paint all parts (adding to the thickness). Do a trial assembly and test for flatness, then make whatever adjustments are necessary.

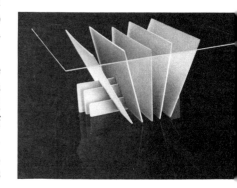

Disassemble the panels for finishing. If you are painting, the color effect can be as austere (all white, say) or as flamboyant as you wish. Once the finish has set, reassemble the ten panels and see how you like the effect, especially with the top in place. If you're not satisfied, you can always repaint some of the panels.

You can, of course, then reassemble the panels permanently, gluing the mating slots. But the beauty of this table is that it is

PANEL LAYOUT

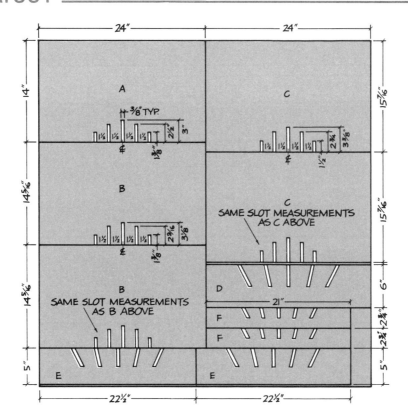

very sturdy without gluing. And this makes it very simple to take apart for flat storage, should the need arise. Who said exceptional beauty and practicality aren't compatible?

MATERIALS LIST

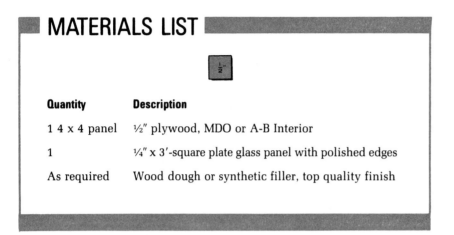

Quantity	Description
1 4 x 4 panel	½" plywood, MDO or A-B Interior
1	¼" x 3'-square plate glass panel with polished edges
As required	Wood dough or synthetic filler, top quality finish

DETAILS OF D, E, & F

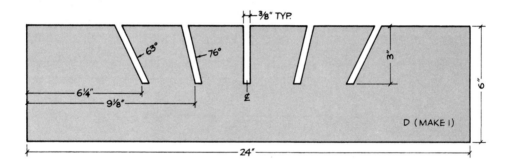

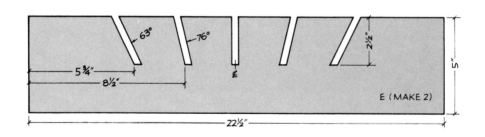

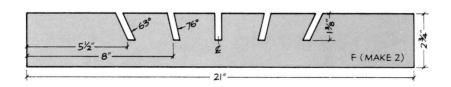

ISOMETRIC VIEW

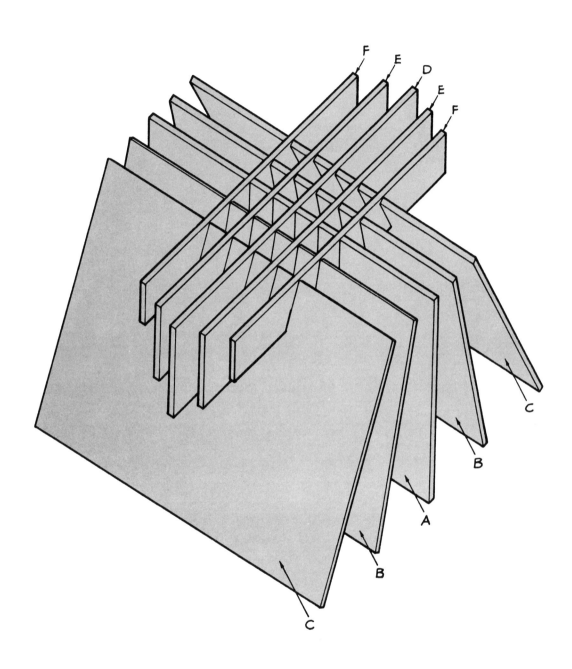

LAMINATED COFFEE TABLE

Even though an important aspect of our design contest is that the project must efficiently utilize the panel or panels required to build it, it's rare that any entry uses every bit of its plywood. This laminated Parsons table requires sectioning the plywood as shown (following the sequence of cuts called for on the next page) and then ripping all pieces into strips so they can be face-glued into the classic Parsons shape.

The designer/builder chose a high-grade panel to minimize voids in the inner plies, since virtually every cut edge will be exposed to view. But even the ends he had left over after the trim cuts were made (dotted lines on the layout sketch) were

Michael Kahney
Lambertville, NJ

ground up into sawdust and mixed with plastic-resin glue to fill what voids there were, before all edges were sanded smooth.

Center saw cuts on layout lines so that pieces are of consistent width (1¾"). When ripping is completed, you should have the following: 28 A strips, 15¾" long; 22 B strips, 14" long; 13 C strips, 66" long; 12 D strips, 62½" long. You miter the ends of two C and four A strips for the edge trim.

When assembling, you may find it easier to laminate flat panels leaving accurate "finger joint" ends, and then join the sections the following day. The mitered edge trim goes on last. You'll

PANEL LAYOUT

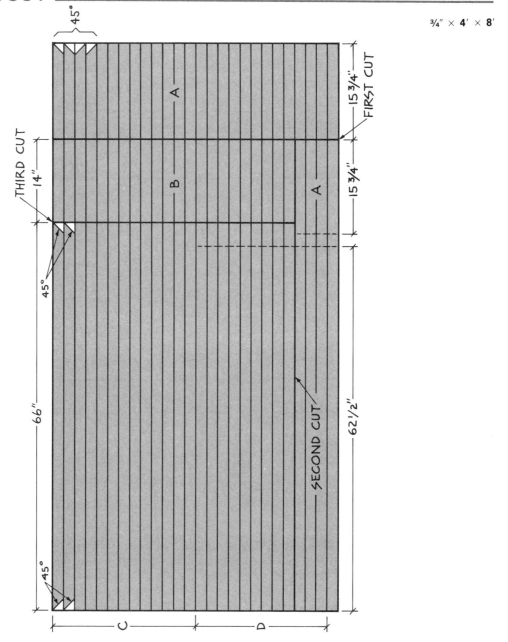

¾" × **4'** × **8'**

need at least two bar clamps to hold assemblies snug while the face-gluing sets. After filling all edge voids, sand all exposed plies smooth, then brush on several coats of clear varnish.

▰ MATERIALS LIST ▰▰▰▰

Quantity	Description
1 4 x 8 panel	¾″ plywood, A-B or A-D Interior, or A-B or A-C Exterior
1 pint (powder)	Plastic resin glue for laminating
1 pint	Satin finish clear varnish (preferably synthetic such as Varathane) compatible with resin glue
As required	Fine sandpaper, wood dough or other wood filler

ISOMETRIC VIEW ═══════════

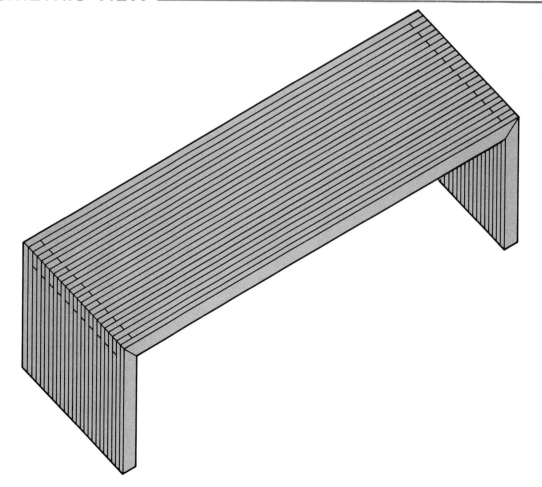

ASSEMBLY DETAIL: CORNER

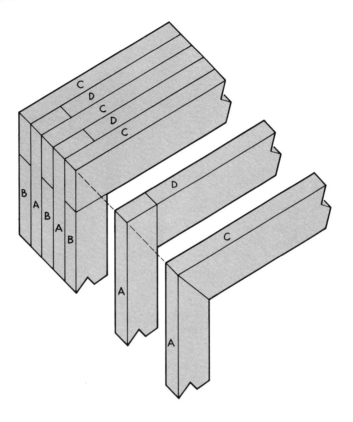

CLAMPING DETAIL

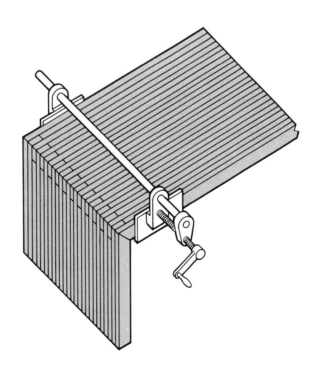

CHILD'S FOLDING TABLE AND STOOLS

Can you believe that this novel folding set of a child's table and four stools was all cut from a half-sheet of ⅝″ plywood? The set is ideal for kids with coloring books, blocks, or even electronic games. And it's perfect for lunch or snacks. Our prototype was two-toned in a mustard yellow and burnt orange for a striped effect. The table stands just under 2′ high and the top is 2′ square. The big bonus is that all five units fold flat for easy portage or storage, as shown in the photo on the next page. This means you can set up for dining in the playroom or on the patio.

Laying out those jigsaw-puzzle parts—which use every square inch of the 4′ × 4′ panel—may be the most time-consuming part of the job. The cutting will go faster than you might think, es-

Harry Oakes
Glen Mills, PA

pecially if you have a large-capacity bandsaw. But first use your table saw (or portable circular saw) to divide the sheet into three pieces: One 2' × 4' and two 1' × 4' pieces. You only have to lay out all those zigzag slats on *one* of these skinny panels if you can tack that one on top of the other and cut both simultaneously on your bandsaw (the 1¼" total thickness would be too much for a portable sabre saw or most jig saws). Center all saw cuts on the layout lines to distribute the kerf widths equally.

Before assembly, however, paint all the zigzag slats and the wooden frame members; it's much easier now than later. Glue-screw the support frames together (forming 11½" squares for the stools, a 22¾" square for the table). Form 1' and 2' squares with the zigzag slats, placing their best sides down on a flat surface. Center the support frame on their backs and attach with finishing nails (or with two 1½" countersunk flathead wood screws through the frame and into each slat).

Now you're ready to attach the leg assemblies. Assuming you want the fold-flat feature, you'll need ten loose-pin hinges and twenty plain butt hinges. Use these to assemble four leg members per unit, as shown on the next page, and attach these five assemblies to the undersides of the stool and table tops. Note that each assembly involves two loose-pin hinges. Replace the pins provided with a nail of the same diameter, bent into an L to make it easier to tug free when you wish to flatten the unit.

PANEL LAYOUT

⅝" × **4'** × **4'**

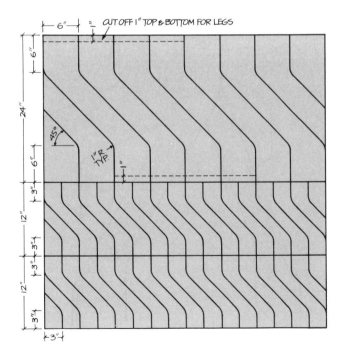

■ MATERIALS LIST ■

Quantity	Description
1 4 x 4 panel	⅝″ plywood, MDO, or, if available, A-B Interior or Exterior
8 lin. ft	1 x 4 framing cut into three 1″-wide strips, then in the following lengths: 21¼″ (2) for table, 22¾″ (2) for table, 9½″ (8) for stools, 11½″ (8) for stools
5 pair	1½″ x 2½″ loose-pin hinges
10 pair	1½″ x 2½″ butt hinges
As required	3d finishing nails, white or urea resin glue, wood dough or synthetic filler, fine sandpaper, top quality latex paint

TABLE: LEGS ASSEMBLY

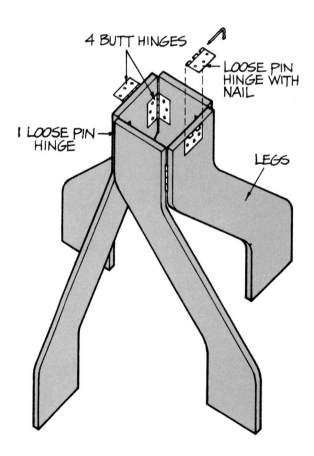

4 BUTT HINGES

LOOSE PIN HINGE WITH NAIL

I LOOSE PIN HINGE

LEGS

TABLE: TOP ASSEMBLY

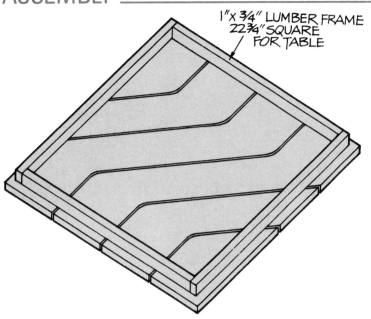

1" x ¾" LUMBER FRAME
22¾" SQUARE
FOR TABLE

TABLE: SIDE & BOTTOM VIEWS

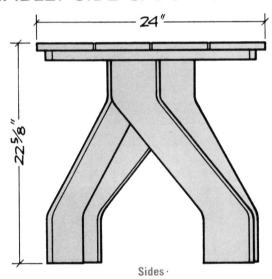

24"

22⅝"

Sides

24"

Bottom

STOOL: SIDE & BOTTOM VIEWS

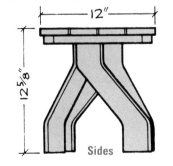

12"

12⅝"

Sides

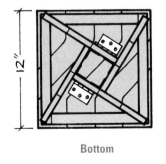

12"

Bottom

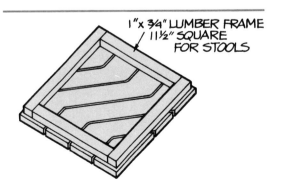

1" x ¾" LUMBER FRAME
11½" SQUARE
FOR STOOLS

ADJUSTABLE LAMINATED CHAIR

One of the most complex entries in seven years of our design competition is this adjustable knockdown chair. All its sections are elaborately laminated from 3"-and 4"-wide planks ripped from two sheets of ¾" plywood. The exposed plies give a striped effect to all edges and faces.

The unit is heavy and sturdy when its panels are assembled with 1" dowels as pivot rods (next page for photo sequence). The back and leg rests can be set at three different angles; the extreme setting aligns them horizontally to provide a flat lounge platform. Toss a pad on this platform and you've an ideal sunning slab—for the patio or for indoor sunlamp use. (Since the chair easily disassembles into portable units, you needn't fret about wrestling

Gareth McAllister
Muncie, IN

the assembled weight from place to place.) And the stacked-lamination system lets you build in conveniences like a swivel-out beverage holder in one arm, a pencil tray (or change caddy) in the other—and clever magazine pouches "sculptured" into both.

Most of the units are laminated of only two ¾" thicknesses, so the laminating process isn't as difficult as it looks. It does get tricky when you get to the arm panels. Study the exploded view and the elevation (which shows details of *both* arm panels in one sketch) to determine exactly how the variously-shaped units stack up to form the magazine pouches and swivel units. Be certain to sand all these special parts thoroughly before gluing. The elevation breaks the stack down into dimensioned units that can be glued up by themselves, using C-clamps. For the final assembly of these units, however, you'll need a pair of pipe or bar clamps.

The most difficult part of this project is the precise drilling of the 1"-diameter holes that align the parts. Also note that the swivel-out units pivot on "captured" dowels that must be glued

PANEL LAYOUTS

³⁄₄" × **4**′ × **8**′

³⁄₄" × **4**′ × **8**′

MATERIALS LIST

Quantity	Description
2 4 x 8 panels	¾″ plywood, A-A, A-B, A-C or A-D
2	1″ x 27″ wood dowel (cut flush with edge of re-taining rings when installed)
2	1″ x 14″ wood dowel (cut flush with edge of leg and back rests when installed)
4	1″ x 2¼″ wood dowel (for leg and back rest reclin-ing supports)
1	¼″ x 1½″ wood dowel (for pencil tray)
1	⅜″ x 3¾″ wood dowel (for glass holder)
As required	White or urea resin glue, wood dough or synthetic filler, sandpaper, top quality finish

into drilled sockets in the planks above and below the units. In the case of the routed tray, the swivel unit is only one lamination high (see the detail for cutting it out of plank 1B); a short ¼″ dowel serves as the pivot. You leave space enough at the other side of the unit to insert a fingertip to start the pivoting action. The beverage holder (with a socket large enough to take a can, bottle or glass) is made up of three drilled thicknesses plus a fourth plank, undrilled, to form the closed base. These all pivot on a ⅜″ dowel, 3¾″ long. You'll have to sand the top and bottom faces of these assembled swivel units enough to prevent binding in their recesses. The magazine pouches form themselves as you stack the variously shaped arm planks.

Note that the 1″ dowels pass through the outer ends of both back and leg rest are permanently glued to the alternating pairs of planks and spacers-disks. The other cross-dowels are free pivots, though the short lengths that capture the anchored end of the notched support arms are glued into sockets in the adjacent planks; this must be done at the same time the end dowels are glued to *all* the members fed into them. The two loose-pivot dowels can have one retaining ring glued to one end. The other ring is drilled for a press fit after the pivots are in place. It is left unglued so you can remove it for disassembly.

Once the assembly is complete, check carefully for any edge voids in exposed plies. Fill these with wood dough, sand all parts carefully and apply several coats of a good polyurethane varnish.

EXPLODED VIEW

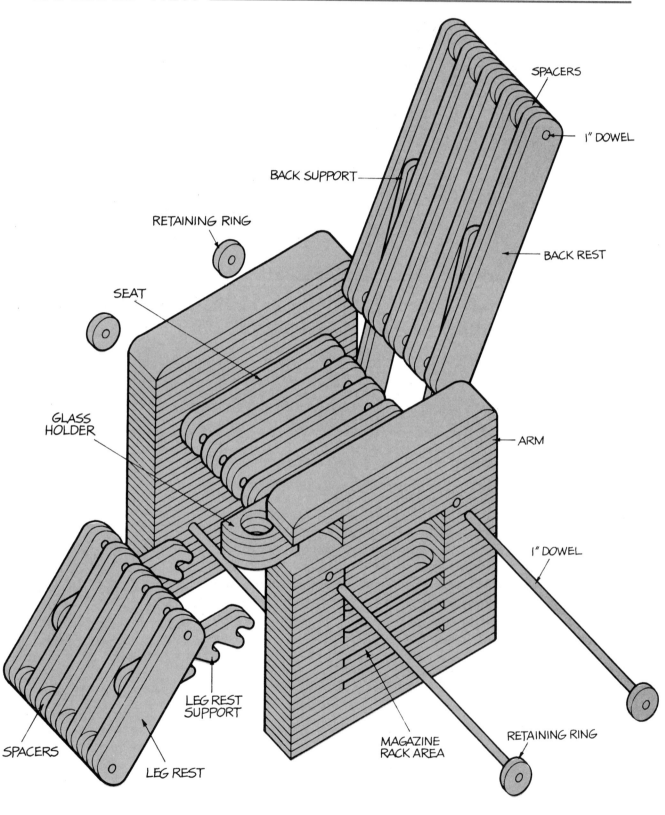

SPACERS

1" DOWEL

BACK SUPPORT

BACK REST

RETAINING RING

SEAT

GLASS
HOLDER

ARM

1" DOWEL

LEG REST
SUPPORT

SPACERS

MAGAZINE
RACK AREA

RETAINING RING

LEG REST

SIDE VIEW

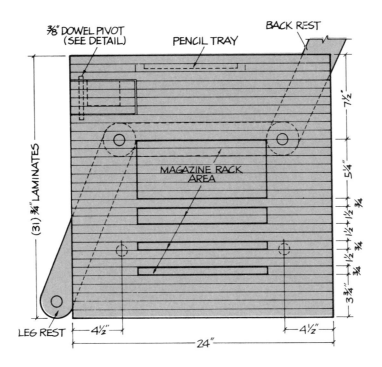

3⁄8" DOWEL PIVOT (SEE DETAIL)

PENCIL TRAY

BACK REST

(31) 3⁄4" LAMINATES

MAGAZINE RACK AREA

LEG REST

7 1⁄2"

5 1⁄4"

3⁄4"

1⁄2" + 1⁄2"

3⁄4"

1⁄2"

3⁄4"

3 3⁄4"

4 1⁄2"

4 1⁄2"

24"

SIDE PIECE DETAILS

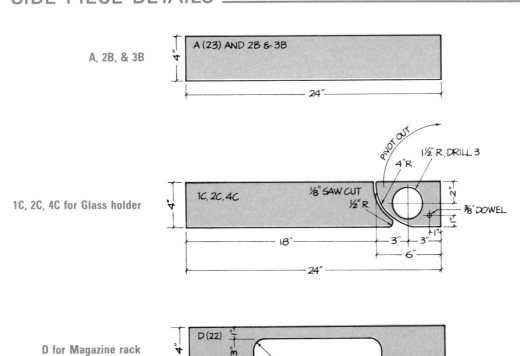

A, 2B, & 3B

A (23) AND 2B & 3B

4"

24"

1C, 2C, 4C for Glass holder

1C, 2C, 4C

PIVOT OUT

1 1⁄2" R., DRILL 3

4" R.

1⁄8" SAW CUT

1⁄2" R.

3⁄8" DOWEL

2"

1"

4"

18"

3"

3"

1"

6"

24"

D for Magazine rack

D (22)

1"

3"

1 1⁄8" R.

4"

6"

12"

6"

24"

MAGAZINE RACK & PENCIL BOX DETAILS

E for Magazine rack

1B for Pencil Box

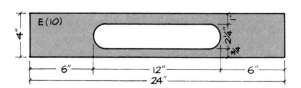

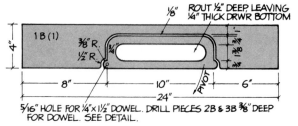

⅛" ROUT ½" DEEP, LEAVING
¼" THICK DRWR BOTTOM

E (10)

4"

6" — 12" — 6"
24"

1B (1)

4"

⅜" R.
½" R.

1"
2¼"
¾"

1⅜"
¾"
¼"

PIVOT

8" — 10" — 6"
24"

5/16" HOLE FOR ¼" x 1½" DOWEL. DRILL PIECES 2B & 3B ⅜" DEEP
FOR DOWEL. SEE DETAIL.

TOP VIEW

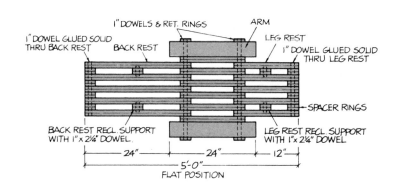

1" DOWELS & RET. RINGS   ARM

1" DOWEL GLUED SOLID
THRU BACK REST   BACK REST   LEG REST

1" DOWEL GLUED SOLID
THRU LEG REST

BACK REST RECL. SUPPORT
WITH 1" x 2¼" DOWEL.

SPACER RINGS

LEG REST RECL. SUPPORT
WITH 1" x 2¼" DOWEL

24" — 24" — 12"
5'-0"
FLAT POSITION

SIDE VIEW

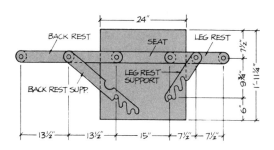

24"

BACK REST   SEAT   LEG REST

7½"

9¾"
1'-1¼"
6"

BACK REST SUPP.

LEG REST
SUPPORT

13½" — 13½" — 15" — 7½" — 7½"

PENCIL BOX PIVOT DETAIL

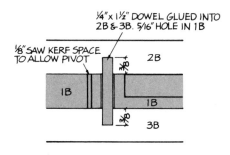

¼" x 1½" DOWEL GLUED INTO
2B & 3B. 5/16" HOLE IN 1B

⅛" SAW KERF SPACE
TO ALLOW PIVOT

2B

⅜"

1B

1B

⅜"

3B

LEG REST & SEAT DETAILS

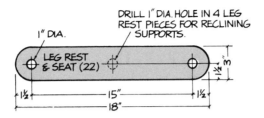

LEG REST SUPPORT DETAIL

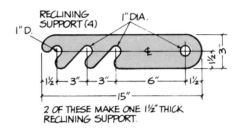

BACK REST DETAIL

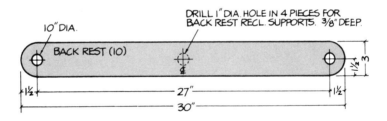

BACK REST SUPPORT DETAIL

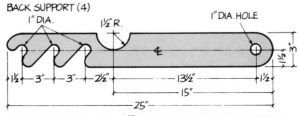

SLOT-TOGETHER ARMCHAIRS

Could you choose between these chairs? I couldn't, while judging the contest for which they were both submitted, so I declared a tie. Each chair efficiently utilizes a half sheet of plywood—so you could build them both from a single 4' × 8' panel. We finished our pair to match two contrasting stripes in our upholstery material; then, when covering our foam cushions we ran the stripes horizontally on one pair of cushions and vertically on the other.

Like other slot-together projects in this book, these chairs can be easily disassembled for flat storage—or easy portage in your car. The more accurately you cut the slots, the more stable the assembled chair will be. It's a good idea to round off each slot's entry edges for easier mating.

David D. Gay
Kalamazoo, MI
and
Thomas Parker
Ypsilanti, MI

When we published these designs as prizewinners in POPULAR SCIENCE, a number of readers wrote to comment on the similarity of the designs to a chair featured in a book titled *Nomadic Furniture*—a redesign of one of the "K-Line" chairs created by Christoph in Germany in 1967. Neither of our prizewinners was a duplicate of that original, however; where you're dealing with such basic simplicity of design, similarities are inevitable.

Our two chairs are quite different. Chair A sits higher and has its back reclining at a greater angle. Note that the height of the slot in each back that takes a tab on the seat is not specified for either chair (though because this slot accepts the seat at an angle, you'll want to make it wider than ¾"). To locate this slot accurately, it's best to cut all parts and slip them together, marking the slot by penciling around the tab as it butts against the back. Disassemble, cut the slot (by drilling a ¾" hole at each end to take the blade of your saber saw), sand all edges and fill any ply voids. The choice of finish depends on where you plan to use the chairs—if outdoors, you'll want to prime and enamel with a good exterior-grade product. You have the option, of course, of a two-tone effect, painting the sides one color and the seat and back as contrasting color.

Try the chairs for comfort, to decide whether you want to upholster them. Cushions are available in a variety of sizes at foam-rubber outlets, if you want to stitch up your own covers.

PANEL LAYOUTS

Chair A   ¾" × 4' × 4'

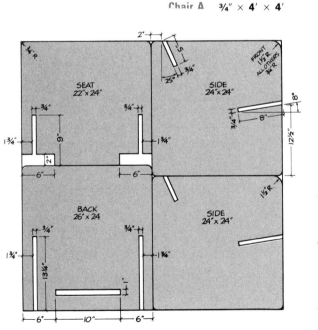

Chair B   ¾" × 4' × 4'

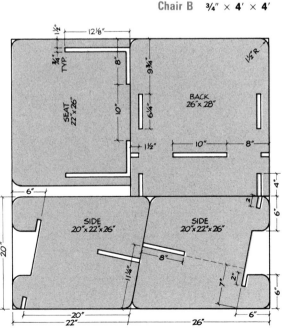

■ MATERIALS LIST

Quantity	Description
1 4 x 4 panel (per chair)	¾″ plywood, MDO or A-B or A-D Interior (for interior use), or A-B or A-C Exterior (for exterior or interior use)
As required	Wood dough or synthetic filler, top quality finish, seat and back cushions (optional)

EXPLODED VIEWS

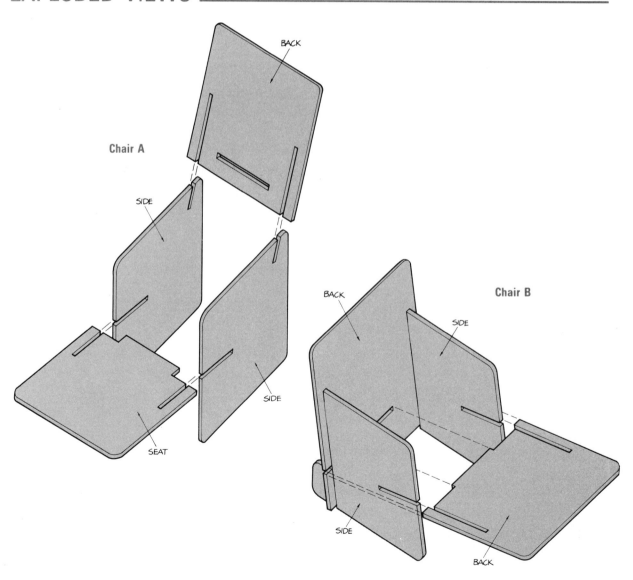

Chair A

BACK

SIDE

SIDE

SEAT

SIDE

Chair B

BACK

SIDE

SIDE

BACK

MODERN ROCKER

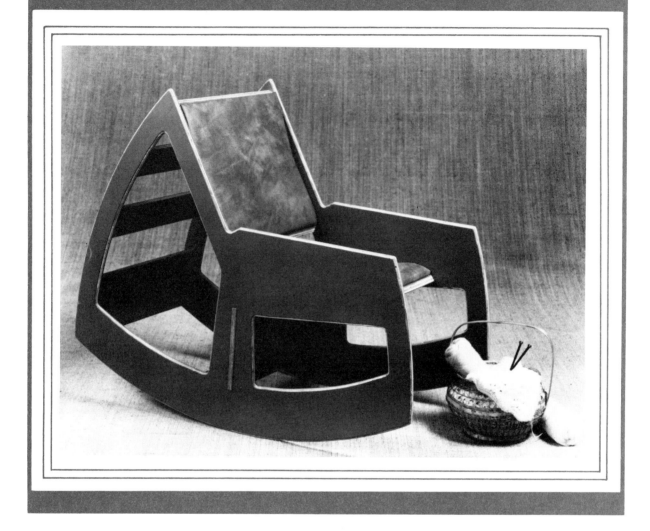

Though this project had more layout waste (on its single sheet of ¾" MDO) than the contest normally permits, we couldn't resist its sleek contemporary lines. And when we built our own prototype from the prizewinning plans, we found it one of the most comfortable chairs we'd ever rocked. For maximum comfort, of course, you'll want to upholster the seat and back. For our rocker, we covered foam padding with suede. With all surfaces painted a matching brown and all edges left natural for contrast, the effect is so striking that everyone who sees the chair wants one like it. (If you'd rather avoid all upholstering, you could pad the chair with loose box cushions available at upholstery outlets.)

Assembly can be with glue and dowels alone, using no screws.

Randy Blake Clontz
Detroit, MI

Either cut up ⁵⁄₁₆″ dowel rods as indicated in the Materials List or buy pre-cut, grooved dowel pegs, an inch long. If you prefer, parts can be screw-glued with flathead screws countersunk and capped with wood dough. Fill any edge voids with the same dough, sand all corners round and coat all edges with two coats of varnish. When that's dry, apply primer and finish coats to all surfaces with a short-nap roller. Doweling permits you to do most finishing *prior* to assembly—a faster and neater scenario.

MATERIALS LIST

Quantity	Description
1 4 x 8 panel	¾″ plywood, MDO or, if available, A-B Interior or Exterior
4′	⁵⁄₁₆″ doweling (for connecting seat and back parts to sides A and B, cut into ⅞″ lengths)
As required	White or urea-resin glue, wood dough or synthetic filler, fine sandpaper, top quality latex paint

PANEL LAYOUT

¾″ × 4′ × 8′

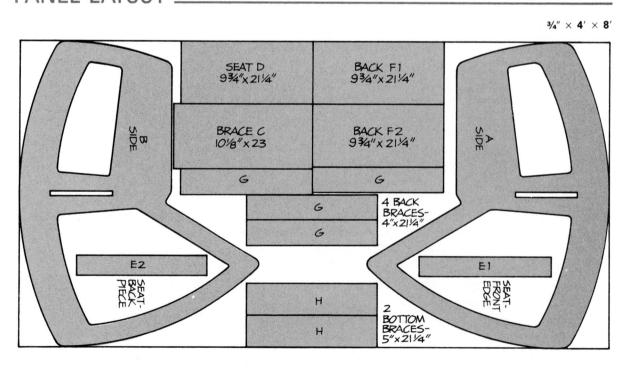

SIDES DETAILS

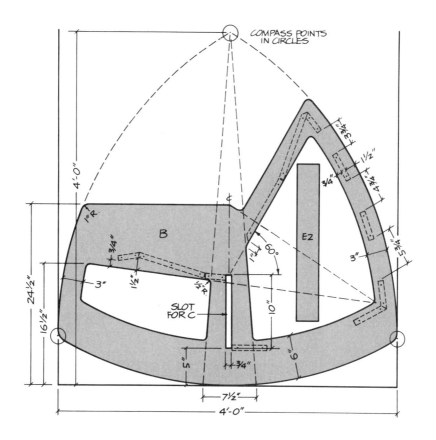

FRONT VIEW

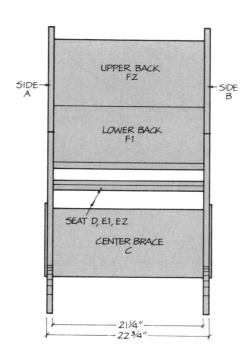

ISOMETRIC VIEW

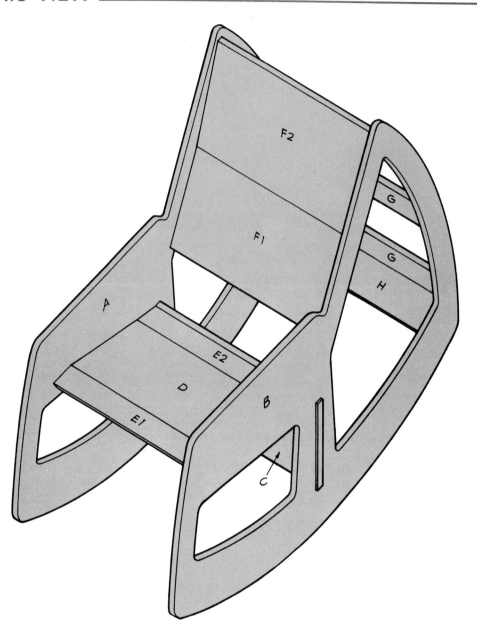

DOWEL & BEVEL DETAILS

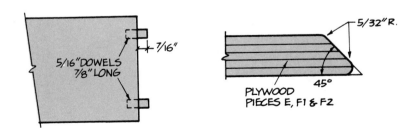

5/16" DOWELS
7/8" LONG

7/16"

5/32" R.

45°

PLYWOOD
PIECES E, F1 & F2

DESKS, DRAWING TABLES, ETC.

MULTI-ANGLE DRAWING TABLE/DESK

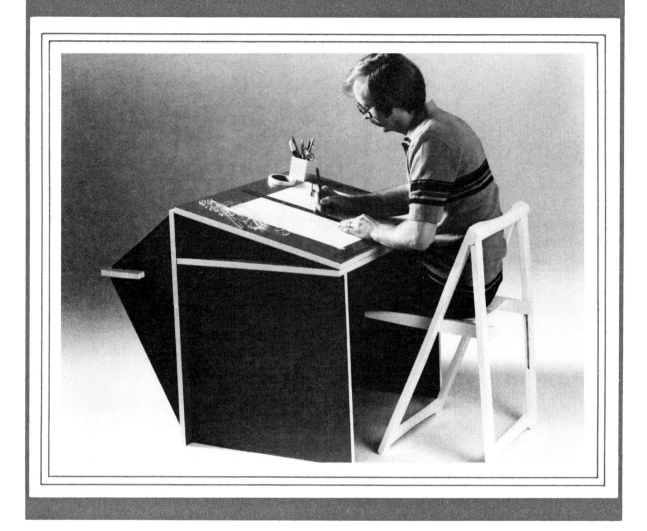

Why did this project win the Grand Prize in our third annual competition? With minimal waste from a single sheet of ⅝" plywood, the designer has created a slot-together work station that can store flat in a closet, yet sets up in a jiffy as a desk or drafting table or an easel.

The three positions of its work surface are achieved by using one of the flat, flanged surfaces (both are the same height) as a standard desk or typing table, then placing an auxilliary top on the angled flanges (photo above) or flipping the whole assembly over to rest on the adjacent square edges so that a steeper angle is brought into play as the support for this auxilliary top (see the "Position 2" sketches on the following pages). There's a bonus:

Gerald R. Herbaugh
APO, NY

83

If you want to keep interrupted work undisturbed or concealed, just lay the angled panel back in place as a cover.

To avoid mechanical fasteners, the designer anchors the angled drawing board with mating Velcro strips on the underside of the panel and along both angles of the supporting edges. For a flat desk top, you just lift off the angled panel when the desk is in either of its positions. You'll note that our Materials List states that the top *can* be held in place by short dowels (¼" dia.) inserted into mating holes drilled into its back surface and into both pairs of supporting edges. The Velcro strips are easier, though, and make for flatter storage.

Actual construction of this project couldn't be simpler. The only thing critical is the spacing of the slots (which should be slightly over ⅝" wide to accommodate panel thicknesses plus surface paint coats, without binding). It's probably best to cut out the basic blanks for both side pieces (A and B) and tack or tape them together so slots can be cut in both simultaneously. This is particularly appropriate if you're working on a bandsaw.

We built our unit of MDO plywood, painted all surfaces a rich green and (after filling all voids) enameled all edges white. One of the easiest ways to achieve this edge treatment is to finish all the edges first, applying two coats and letting them dry thoroughly. Then, roller-coat all surfaces with a primer and a finish coat, taking care not to let sags or runs drip down the edges. If drips *do* occur, wipe them away with a clean cloth (moistened in paint thinner, if necessary).

PANEL LAYOUT

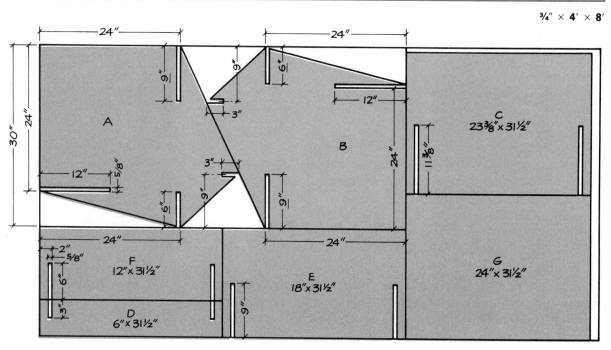

¾" × **4'** × **8'**

MATERIALS LIST

Quantity	Description
1 4 x 8 panel	⅝″ plywood, MDO, A-C Exterior, or, if available, A-B Interior
As required	Wood dough or synthetic filler, fine sandpaper, top quality latex paint

NOTE: The angled drawing surface (G) can be held in place by dowels inserted into the top edges of sides A and the bottom of G, or by non-permanent-stick adhesive strips, such as Velcro tape. (Velcro consists of two nylon mesh tapes which when pressed together form a secure but releasable bond.)

POSITION 2: SIDE, TOP, & FRONT VIEWS

Top

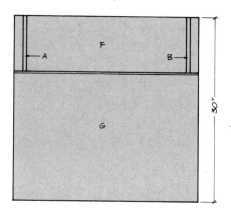

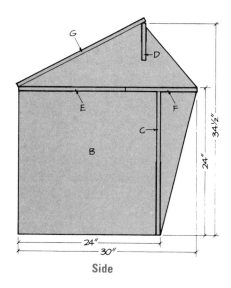

Side

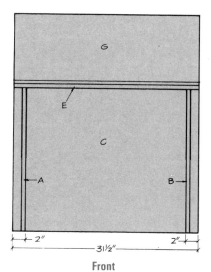

Front

POSITION 2: ISOMETRIC VIEW

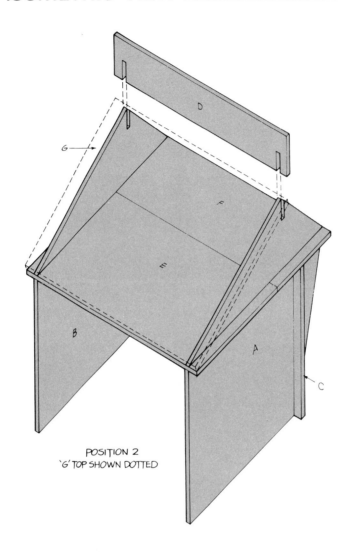

POSITION 2
`G' TOP SHOWN DOTTED

EXPLODED VIEWS

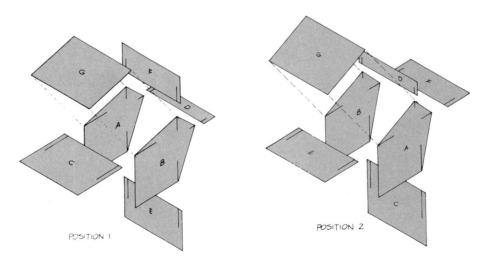

POSITION 1

POSITION 2

POSITION 1: ISOMETRIC VIEWS

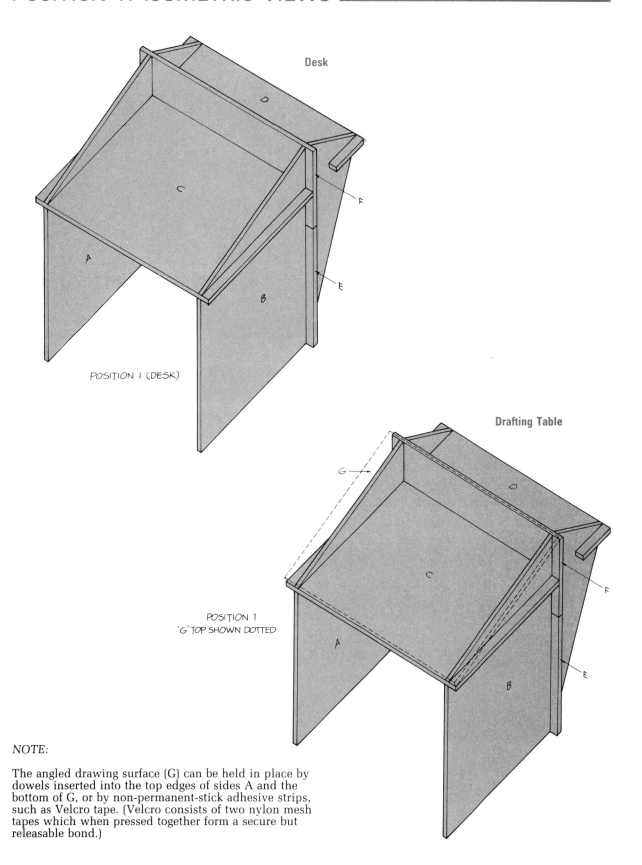

Desk

POSITION I (DESK)

Drafting Table

POSITION 1
'G' TOP SHOWN DOTTED

NOTE:

The angled drawing surface (G) can be held in place by dowels inserted into the top edges of sides A and the bottom of G, or by non-permanent-stick adhesive strips, such as Velcro tape. (Velcro consists of two nylon mesh tapes which when pressed together form a secure but releasable bond.)

KNOCKDOWN HOBBY TABLE

It bolts and slots together into a sturdy work table with a 24″ × 40″ top, yet you can get all the pieces out of a half-sheet of ¾″ plywood, with *no* waste except for what you cut out to form the ¾″-wide slots.

The top has drop flanges glued and screwed along its two sides, and these are drilled to take a pair of round-head stove bolts through the leg uprights. By drilling a series of paired holes through the uprights, as shown in the assembly drawings (turn page), you can easily adjust the height of the work surface to fit the work and the worker: Bolt the top at 29″ for normal desk work, 27″ for your home computer or 25″ for use by a child. Underslung drawers are an option (not provided for here).

Luis E. Arroyave
Guatemala City, Guatemala

When you want to take the desk with you, just back out the bolts, tap up the cross-brace to slide it free of the leg slots, and all parts go into the short stack shown in the photo below.

Since we planned to enamel our desk in two contrasting tones (the top assembly and the cross-brace a burnt orange, the legs a suede brown) we built it out of MDO. It's a super-simple project you can complete, ready for use, in a weekend. A portable circular saw will speed initial cutting, and if you use a plywood blade you'll minimize edge-sanding. Slice off the cross-brace plank first, then cut the top panel from the remaining sheet. Next, cut the two side-flange planks, then use a sabre saw to cut apart the nested leg units—and to cut the four slots. You'll need a ¾″ drill bit

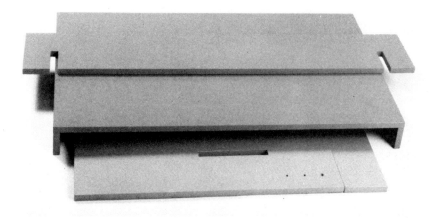

PANEL LAYOUT

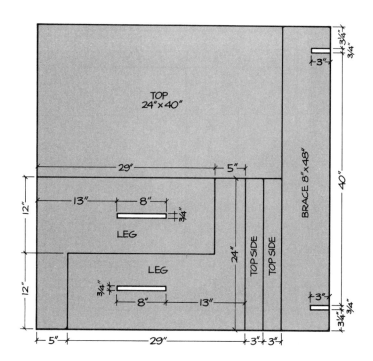

to pierce the panel at the end of each slot. For the slots in the legs, just drop the blade through one of the holes and cut toward the other one. (It's best to clamp a straight-edge against the surface to guide the saw base.) It's easy to square up the corners after the scrap piece falls free. Smooth and fill all edges before applying your prime coat.

MATERIALS LIST

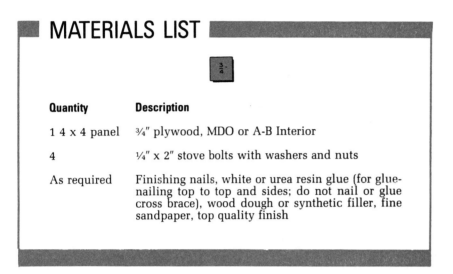

Quantity	Description
1 4 x 4 panel	¾″ plywood, MDO or A-B Interior
4	¼″ x 2″ stove bolts with washers and nuts
As required	Finishing nails, white or urea resin glue (for glue-nailing top to top and sides; do not nail or glue cross brace), wood dough or synthetic filler, fine sandpaper, top quality finish

SIDE VIEW

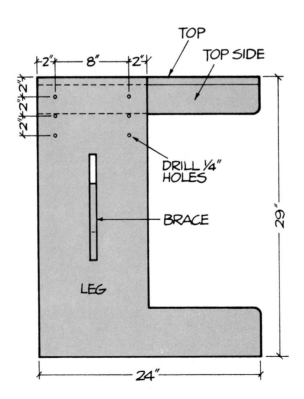

EXPLODED VIEW

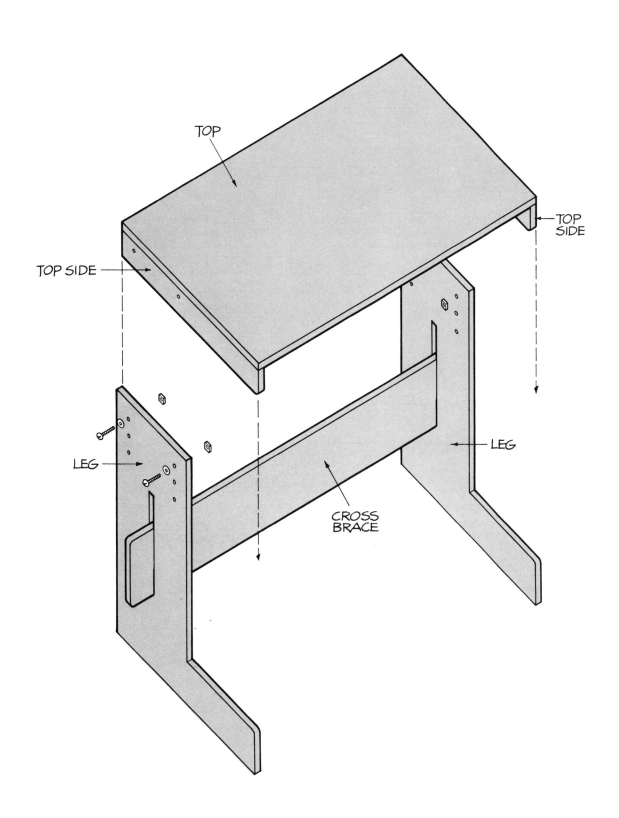

TOP

TOP SIDE

TOP SIDE

LEG

LEG

CROSS BRACE

TWO-POSITION DESK/DRAFTING TABLE

Yet another variation on the slot-together, store-flat, adjustable-top desk, this clever unit is cut from a single sheet of ¾″ plywood with no waste except for the cut-out slots. It sets up in minutes into a sturdy desk with a work surface just under 28″ off the floor—a compromise height that serves both for study and typing.

But pull up the top rail (which, in this position, has mating equal-depth slots that set it flush with the tops of the end panels it bridges), invert it, and the rail now rides well above the sides to tip the work surface at a proper angle for a drafting table (see smaller photo, next page). The edge that now engages the slots in the ends and the divider has only shallow notches to hold it in position. What has now become the rail's *top* edge has two

John Leggitt
Santa Barbara, CA

dowel pegs protruding ½″ to engage a pair of oversized holes in the underside of the work top. Note that similar dowels, glued in the upper edges of the end panels, engage four holes at the corners of the work top when the top is in its flat position.

This project boasts two economies: It utilizes its panel to the fullest and requires no hardware. Assembly is without glue or screws. It's best to stack the two end panels to cut their vertical slots at the same time. Separate the panels and stack the right end and divider panel to cut their matching horizontal slots. The two identical shelves can be stacked and cut to shape simultaneously, as well. (The shelf compartment has no back, although the top rail closes off the top shelf when it's in its lower position.)

The easiest way to cut the slots is to drill a ¾″ hole at the end of each and then saw into both sides of these holes along the layout lines. A bandsaw will give you the straightest sides, especially if you're cutting stacked panels, but you can manage the cuts with a wide-bladed sabre saw. As with the other cuts, center the kerf on the layout line: This will make your slots slightly wider than the panel thickness, to accommodate the finish coats on each surface.

We enameled our prototype a single color, painting the folding chair to match, but a contrasting edge treatment would add a design element. Be sure to fill all edge voids, sand smooth and slightly round corners before applying the prime coat.

PANEL LAYOUT

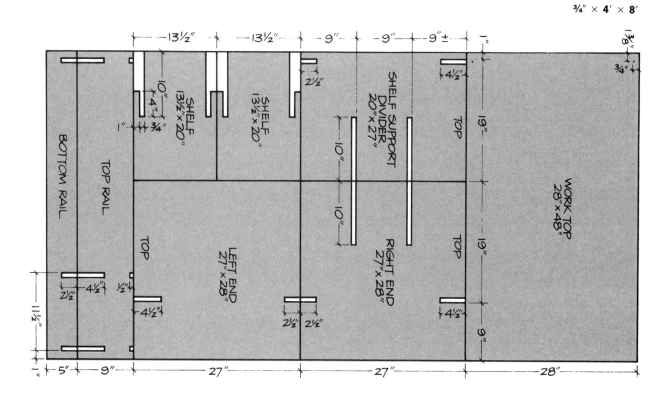

MATERIALS LIST

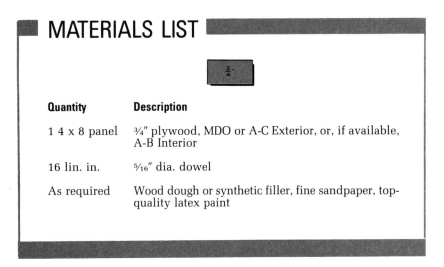

Quantity	Description
1 4 x 8 panel	¾″ plywood, MDO or A-C Exterior, or, if available, A-B Interior
16 lin. in.	⁵⁄₁₆″ dia. dowel
As required	Wood dough or synthetic filler, fine sandpaper, top-quality latex paint

ISOMETRIC VIEW

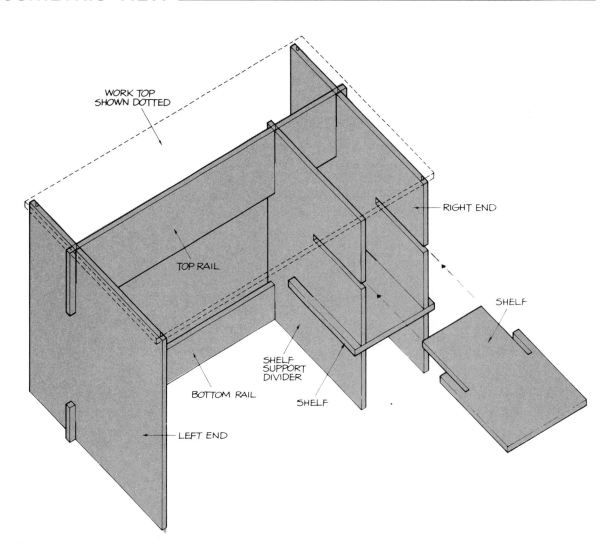

WORK TOP
SHOWN DOTTED

RIGHT END

TOP RAIL

SHELF

SHELF
SUPPORT
DIVIDER

BOTTOM RAIL

SHELF

LEFT END

FRONT VIEW

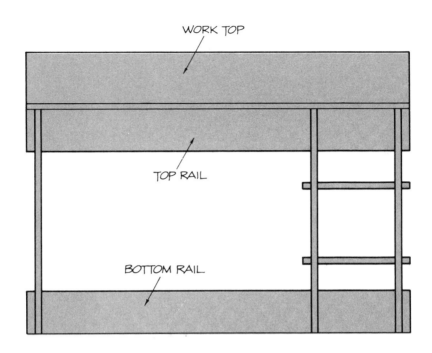

WORK TOP

TOP RAIL

BOTTOM RAIL

SIDE VIEW & DOWEL DETAIL

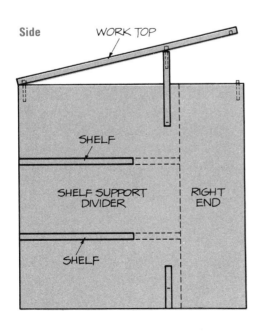

Side

WORK TOP

SHELF

SHELF SUPPORT DIVIDER

RIGHT END

SHELF

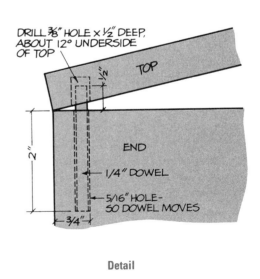

DRILL ⅜" HOLE x ½" DEEP, ABOUT 12° UNDERSIDE OF TOP

½"

TOP

END

2"

1/4" DOWEL

5/16" HOLE-
SO DOWEL MOVES

3/4"

Detail

ART AND DRAFTING DESK

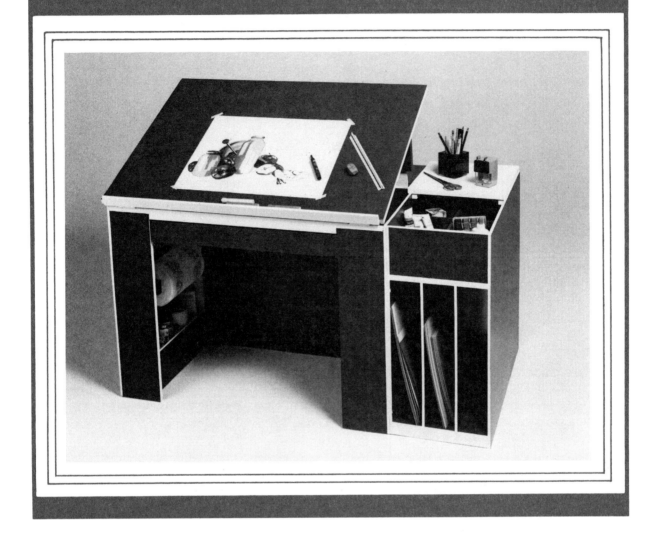

At the time this project was submitted its designer was an art student. While working on a painting at home during a school break, she grew frustrated with the confused tumble of her art supplies and planned a storage organizer with a big, hinged desk top that could be set at several angles. She has since moved it often: The unit disassembles easily.

Construction requires three 4' × 8' sheets of plywood—one ¾" thick and two ⅜" thick. The final assembly is "two-faced"– as the photos show. There are many storage compartments, including tall slots for art boards and canvases. There's even a rack for a roll of paper towels tucked in the knee well. The four-position top goes from flat (for a large desk or work table) to ever

Mary Elaine Buchholtz
Ames, IA

steeper angles set by adjustable supports. At 15 degrees it's a drafting table; at 85 degrees it's an artist's easel.

With as complex a project as this, letter all parts on the layout. They'll be easier to keep track of once they're cut. First assemble the upright side supports, glue-nailing all joints. (That involves ¾" parts J, K, L, M and P, plus ⅜" parts C, G and R.) Glue-nail front-tie strip B to the center of strip H, then screw this tie across the two support assemblies, using the four 1½"-long flatheads. Screw on the 3" pair of loose-pin hinges, centered between the first screwheads, as shown in the first assembly sketch. Glue and nail the drawer together (from ¾" part F and ⅜" parts J, K, U, V and W). Assemble the bookcase unit, glue-nailing all joints (¾" parts G and I; ⅜" parts D, H, I, L and N). Attach doors with hinges and catches.

Cut two dadoes and a rabbet the full length of the storage unit base and top (¾" parts C); assemble with three partitions and the outside panel (⅜" parts A and B). Attach kick strips (¾" parts O), then assemble the cubbyhole tray (¾" ends E plus egg-crate-slotted ⅜" parts P and Q). Glue-nail position strips to underside of storage unit lid (¾" parts N and D). Attach the bookcase and storage units to the upright support assembly with the roundhead screws and finish washers—three on each end of the bookcase, two on front of storage, three on back.

Glue-nail the edge stop (⅜" part M) to the drawing board (¾" part A), assemble the support holders (⅜" parts E and F) and hinge these to the back of the drawing board. Attach the board to the hinges on the tie strip. Cut the safety chain to a length that will prevent the board from flopping too far forward. The three pairs of support bars (⅜" parts S, T and O) hold the drawing board at its three angles.

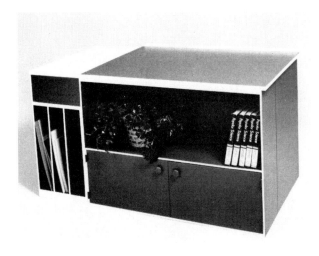 

PANEL LAYOUTS

$\frac{3}{8}'' \times 4' \times 8'$

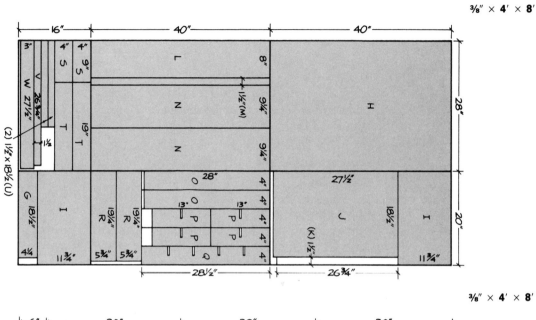

$\frac{3}{8}'' \times 4' \times 8'$

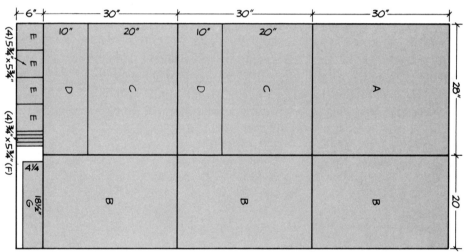

$\frac{3}{4}'' \times 4' \times 8'$

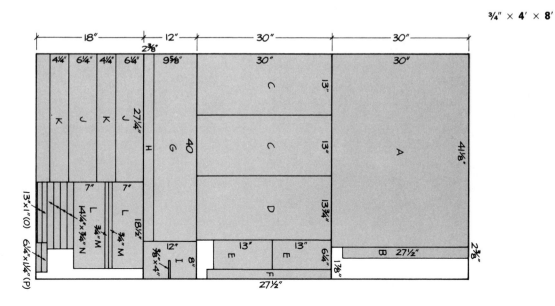

▪ MATERIALS LIST

Quantity	Description
1 4 x 8 panel	¾″ plywood, MDO or, if available, A-B Interior
2 4 x 8 panels	⅜″ plywood, similar grade
2 pr. each	Loose-pin cabinet hinges, cabinet hinges, 3″ x 1″
4 pr.	Cabinet hinges, 2″ x 1″
2	Cabinet door catches
60 lin. in.	Safety chain (Approx. length)
4	Small screw eyes
4	1½″ x ¼″ #20 flathead wood screws
11	¾″ x ¼″ #20 roundhead wood screws with finish washers
As required	Finishing nails, white or urea resin glue, fine sandpaper, top-quality enamel and companion primer

BOOKCASE

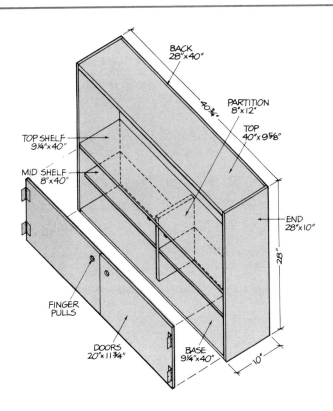

STORAGE UNIT

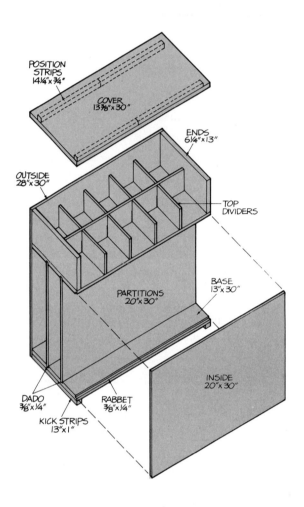

POSITION
STRIPS
14¼" x ¾"

COVER
13⅜" x 30"

ENDS
6¼" x 13"

OUTSIDE
28" x 30"

TOP
DIVIDERS

BASE
13" x 30"

PARTITIONS
20" x 30"

INSIDE
20" x 30"

DADO
⅜" x ¼"

RABBET
⅜" x ¼"

KICK STRIPS
13" x 1"

DRAWER

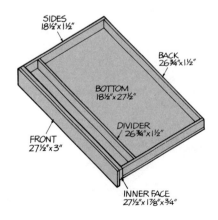

SIDES
18½" x 1½"

BACK
26¾" x 1½"

BOTTOM
18½" x 27½"

DIVIDER
26¾" x 1½"

FRONT
27½" x 3"

INNER FACE
27½" x 1⅞" x ¾"

UPRIGHT SUPPORTS

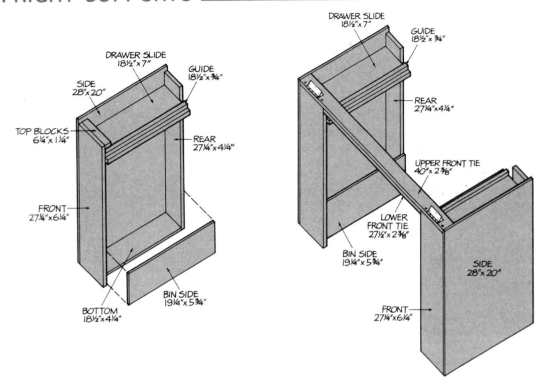

DRAWER SLIDE
18½"x 7"

SIDE
28"x 20"

GUIDE
18½"x ¾"

TOP BLOCKS
6¼"x 1¼"

REAR
27¼"x 4¼"

FRONT
27¼"x 6¼"

BOTTOM
18½"x 4¼"

BIN SIDE
19¼"x 5¾"

DRAWER SLIDE
18½"x 7"

GUIDE
18½"x ¾"

REAR
27¼"x 4¼"

UPPER FRONT TIE
40"x 2⅜"

LOWER
FRONT TIE
27½"x 2⅜"

BIN SIDE
19¼"x 5¾"

SIDE
28"x 20"

FRONT
27¼"x 6¼"

DRAWING BOARD SUPPORT DETAILS

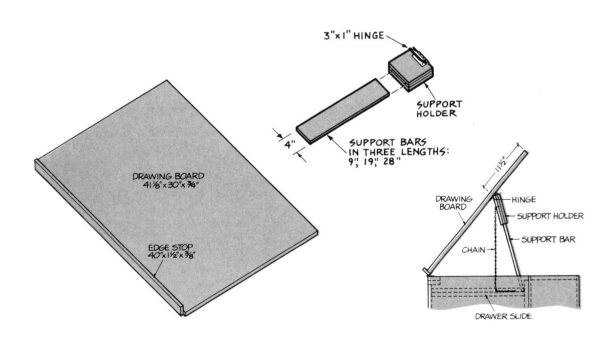

3"x1" HINGE

SUPPORT
HOLDER

4"

SUPPORT BARS
IN THREE LENGTHS:
9", 19", 28"

DRAWING BOARD
41⅛"x 30"x ¾"

EDGE STOP
40"x 1½"x ⅜"

11½"

DRAWING
BOARD

HINGE

SUPPORT HOLDER

SUPPORT BAR

CHAIN

DRAWER SLIDE

SECRETARY WITH CHAIR

Designed to fit any spare nook in your home, this compact desk is ideal for those household management chores—compiling shopping lists, balancing checkbooks, scheduling activities, dashing off short personal notes. And this desk and chair would make a fine addition to a teen-ager's room, as well. Both the 5'-tall secretary and the pedestal chair come neatly out of a single sheet of ⅝" plywood.

A drop-leaf writing surface provides ample space and lends a tidy appearance when folded up at the completion of the work. Three tall shelves and a handy compartment just under the hinged leaf offer storage space for books, ledgers, telephone directories and display items.

Michael N. Ward
Tulsa, OK

Aside from the hinges, you'll need a pair of sliding-pivot lid supports—preferably in a brass to match the hinges. And if you want the shelves to be adjustable, as we did, you just drill a series of ½" holes, 1" o.c. and about ⅜" deep into the inner faces of the two side members. After assembly, just tap in four 1¼"-long dowels at each desired height to support each shelf. (If you prefer, of course, you can glue and screw the shelves at permanent locations.)

As the assembly diagram shows, most joinery is done with simple butt joints, glued and screwed. But the cubbyhole section inside the desk should be set into routed grooves, as indicated on the cutting diagram. Note that the fixed panel of the desk top is also supported on 1 × 1 cleats (or a 1" quarter-round strip). For a bit of extra trim, the hinged lid is routed with a cove bit.

The cross-brace at floor level is butt-joined with glue and two finishing nails into each end. All nail and screw heads should be driven beneath the surface and filled with putty. Fill any edge voids at the same time and sand smooth, rounding all edges slightly.

The unit, as shown, has no back, so you'll want to finish it to compliment or contrast with the color of the wall it's to be set against. If you plan to paint the unit, as we did, you should build it from an MDO panel. We enameled ours solid blue, but this project could benefit from a contrasting edge treatment. You might even consider finishing the chair to match the desk's edge-color.

PANEL LAYOUT

⅝" × **4'** × **8'**

■ MATERIALS LIST ■

$\frac{5}{8}''$

Quantity	Description
1 4 x 8 panel	⅝″ plywood, MDO or A-B Interior
2	2″ brass hinges with screws
2	10″ brass lid supports with screws
8	½″ x 1¼″ dowels
2	1″ x 1″ x 15″ lumber desk top support cleats
2	1″ x 1″ x 13″ lumber chair seat support cleats
As required	Finishing nails, white or urea resin glue, wood dough or synthetic filler, fine sandpaper, top quality finish

ISOMETRIC VIEW

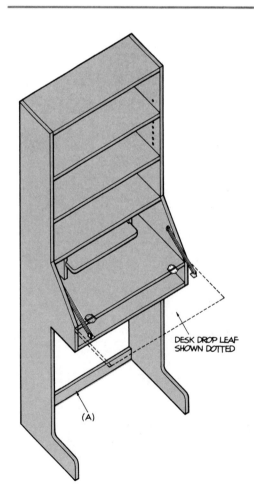

DESK DROP LEAF
SHOWN DOTTED

(A)

FRONT & SIDE VIEWS

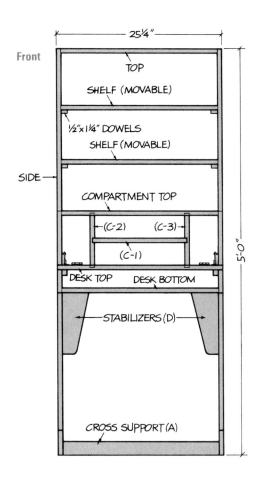

25¼"

Front

TOP

SHELF (MOVABLE)

½"x1¼" DOWELS

SHELF (MOVABLE)

SIDE

COMPARTMENT TOP

(C-2)   (C-3)

(C-1)

DESK TOP   DESK BOTTOM

STABILIZERS (D)

5'-0"

CROSS SUPPORT (A)

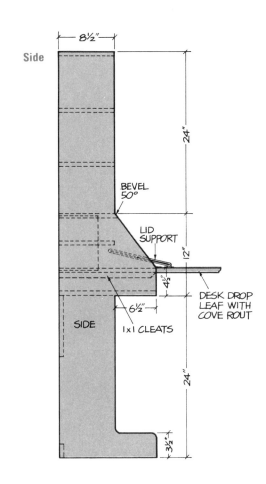

8½"

Side

24"

BEVEL 50°

LID SUPPORT

12"

4½"

6½"

SIDE

1x1 CLEATS

DESK DROP LEAF WITH COVE ROUT

24"

3½"

CHAIR VIEWS

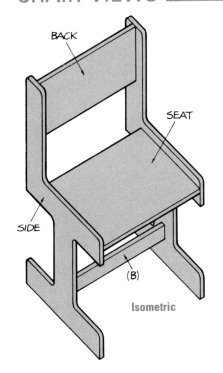

BACK

SEAT

SIDE

(B)

Isometric

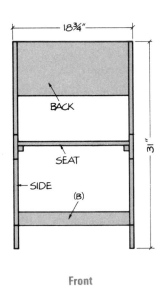

18¾"

BACK

SEAT

SIDE

(B)

31"

Front

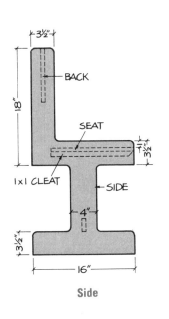

3½"

BACK

SEAT

18"

1x1 CLEAT

SIDE

3½"

4½"
3½"

4"

16"

Side

SECRETARY/VANITY WITH BENCH

Whether you build this as a desk or an accounts center for the kitchen, add a mirror to create a teenager's vanity, or use it as a combination shelf/writing desk for a guest room, you can't make better use of a single sheet of ¾″ plywood. Note from the layout that the cutouts in the desk sides become the legs of the bench. To scribe the 13½″ radius for these, you'll need a pencil on a string or—for better accuracy—a bar compass made from a 14″ length of lath with a hole drilled ¼″ from each end. You spike one end with a pin and insert your pencil point through the other hole to scribe the arc. The 2″ radius for many of the corners is best scribed with a simple school compass.

Since there's no waste area, you must make the bench-side cut-

Richard Guimond
Bowie, MD

outs with special care—starting them with a cautious plunge cut or by inserting the blade through a slot formed by drilling a series of small holes along the cutting line. When cutting is complete, smooth all edges with sandpaper.

Drill pilot holes and fasten backstops to shelves with glue and screws. In the same way, attach the desk-top front strip and the two ledgers that support the top. Mark locations for the shelves and the cross brace and drill clearance holes for screws through the uprights. Assemble with glue and screws.

To assemble the bench, first join the four seat supports to form a shallow box with their top edges flush. Screw and glue the three top slats to these edges. Attach the sides and the rails by driving screws through the side panels.

Since so many screws are involved, it's easiest to use the type with oval heads so you can leave them exposed. If you use flat-head screws, countersink them and cover the heads with wood dough or plugs. Fill edge voids at the same time, sand smooth, and prime for painting. Our prototype was painted a mustard yellow, but the unit would be even more striking if the exposed edges were enameled a contrasting color. The bench could then be painted either to duplicate the two-tone design of the desk, or solid in the edge-color used for the desk.

PANEL LAYOUT

¾" × 4' × 8'

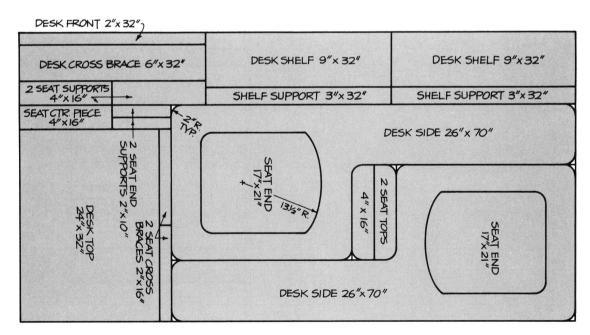

▇ MATERIALS LIST ▇

Quantity	Description
1 4 x 8 panel	¾" plywood, A-B or A-D Interior, A-B or A-C Exterior, or MDO
2 pieces	1 x 2 lumber 24" long for desk supports
40 approx.	#10 x 2" oval or flathead wood screws for glue-screw construction
8 to 10	3d casing or finishing nails for glue-nailing ledger strips
As required	White glue for glue-screw construction; wood dough for filling screw holes and any small gaps in plywood cut edges; fine sandpaper for smoothing cut edges and cured wood dough; finishing materials: primer and paint, antiquing kit, or synthetic satin-finish varnish (e.g., Varathane) with or without stain

SIDE & FRONT VIEWS

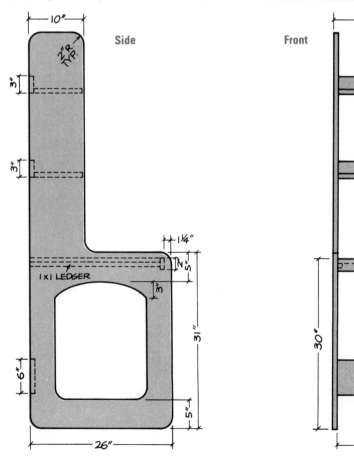

Side

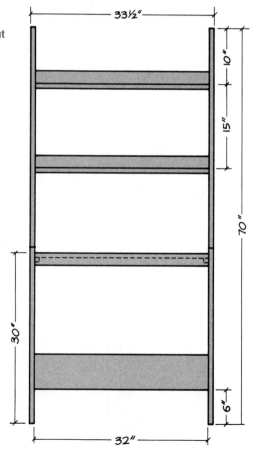

Front

EXPLODED VIEW

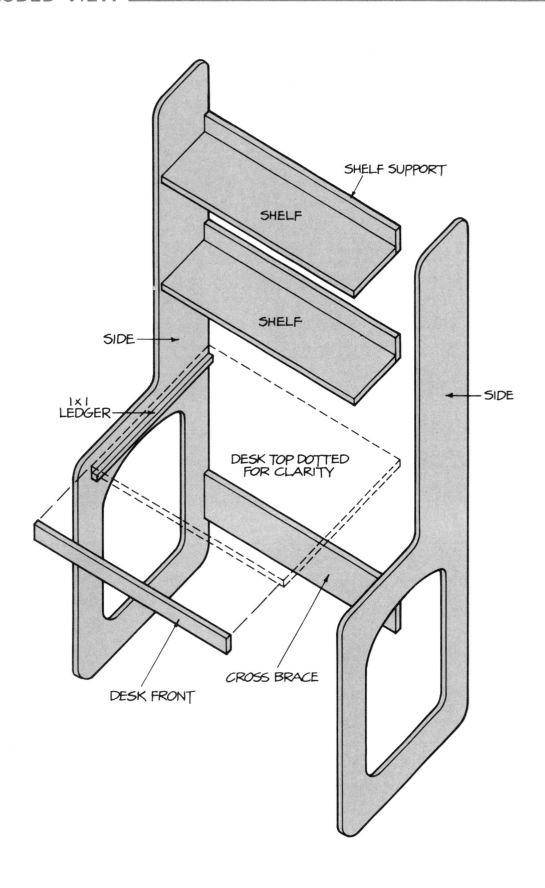

SHELF SUPPORT

SHELF

SHELF

SIDE

1 x 1
LEDGER

SIDE

DESK TOP DOTTED
FOR CLARITY

CROSS BRACE

DESK FRONT

BENCH VIEWS

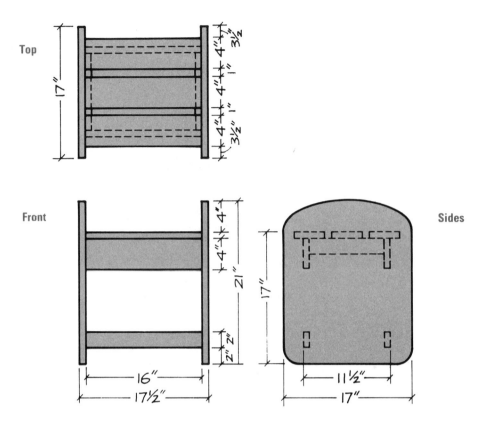

Top

17"

3½
4"
4"
1"
4"
1"
4"
3½

Front

4"
4"
4"
21"
2"
2"

16"

17½"

Sides

17"

11½"

17"

BENCH EXPLODED VIEW

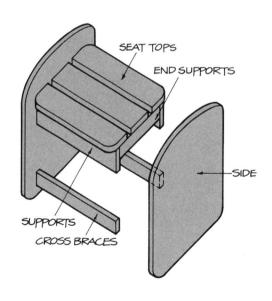

SEAT TOPS

END SUPPORTS

SIDE

SUPPORTS

CROSS BRACES

CUBE DESK

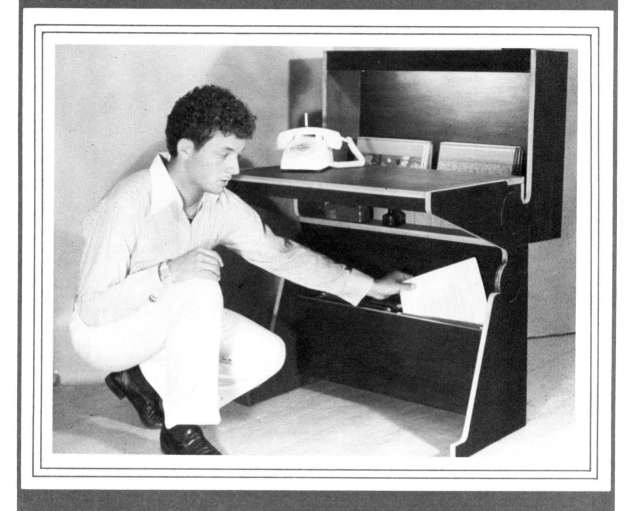

Here's a unique design to test your woodworking skills. The desk folds into a cube (as shown in photos on the next page) for storing or moving, and the pivots are built into the design. They consist of lollipop extensions of the plywood sides that are locked into mating cavities by means of flexible rods set into grooves in the mating edges to form a track.

Before attempting the panel layout for cutting the pieces, you'll want to prepare a full-size template of these disk pivots from the dimensioned drawing on page 116. (And when you come to cutting apart the mating members, bear in mind that the designer/builder laboriously made these cuts with a fine-blade coping saw that

George S. Kanelba
New York, NY

PANEL LAYOUT

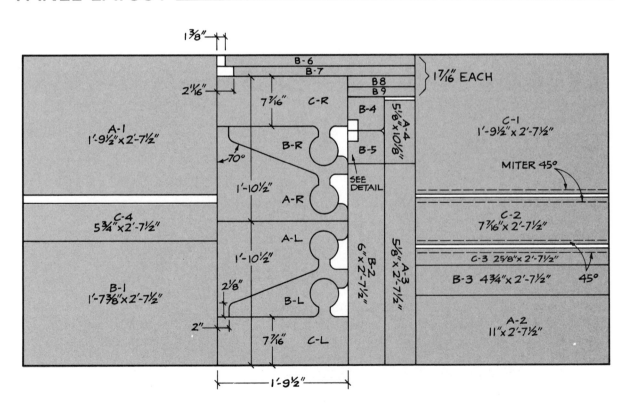

left only a ⅟₁₆″ kerf.) Reverse the template when you are tracing the layout for the right-side pivots.

Make all primary cuts with a standard plywood or combination blade leaving a saw kerf of ⅛″, leaving the critical mating hinge cuts until last. Then you must cut ⁵⁄₁₆″-dia. grooves along the centerlines of the edges of all mating parts. The best method is to use a ball-shaped, high-speed steel cutter bit (Dremel No. 114) chucked into a drill press. Set the bit to center ⅜″ above the drill-press platen, and move the work against the bit. Next, drill ⁹⁄₃₂″ dia. holes, ⅜″ deep, at the ends of the grooves in each male hinge part (see template drawing).

To conceal the wire to the built-in strip fixture in the top compartment, cut a channel down one vertical edge of Panel *C-1*. Install the wire before glue-nailing the side pieces to this back

■ MATERIALS LIST ■

Quantity	Description
1 4 x 8 panel	¾″ plywood, A-A or A-B Interior, A-B or A-C Exterior, or MDO
1 piece	Hardboard, 32¾″ x 11⅛″ x ⅛″ for drawer bottom
1	Cylinder lock for drawer
1 set	Drawer glides, 12″ long, Grant A-13-9 record player slides or equivalent (or see alternate drawer glide detail)
1 length	Smooth-coated cable (rubber- or plastic-coated hi-fi coaxial, low-voltage is fine), 6′ x ¼″ dia. for plywood hinge edges
1	Small tube graphite powder for hinge cable
1	Lighting strip, miniature fluorescent or incandescent (e.g. C.J. Lighting Co., Inc., No. R701 using four T-10 finger bulbs) with cord as required and plug
4	Steel bars, ⅛″ x ½″ x 5⅞″ long for hanging file folders (Pendaflex type)
1 set	Furniture glides (or levelers if needed)
1 piece (optional)	Plastic laminate, 19⅜″ x 31½″ for desk surface (cut Panels B-L and B-R ⅟₁₆″ short to accommodate thickness)
As required	4d finishing nails and white glue for glue-nail assembly; filler for countersunk nail holes and edge voids; fine sandpaper; finishing materials

DETAILS: DESK PIECES B-4 & B-5

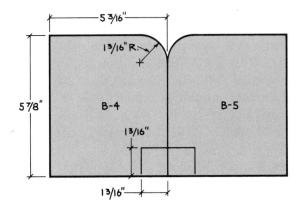

SIDE VIEW

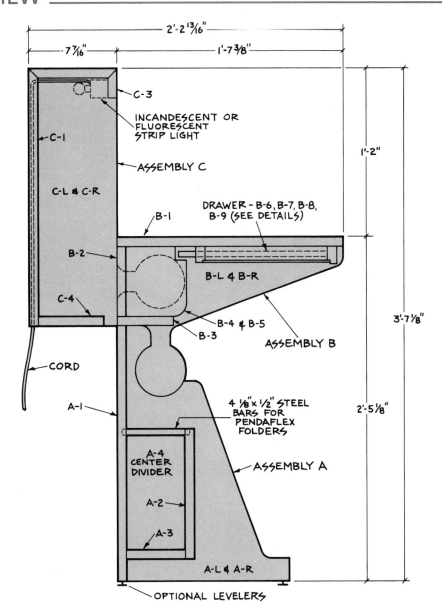

panel, taking care not to nail through the electrical wire. Glue-nail each assembly separately, as follows: Assembly A: First cut slots in panels A-1 and A-2 for file hanger bars. Assembly B: Glue-nail panels B-L, B-R and B-3 (B-1 is installed after the desk is assembled). Glue-nail panels B-2, B-4 and B-5 together and set aside. You can assemble and install the drawer after the entire desk is assembled.

Apply finish to all assemblies. A dark stain was used on the prototype shown, with the edges highlighted in a white enamel. White enamel was also applied to the inside faces of panels C-2 and C-3 for light reflectance.

Now, cut your chosen rod or cable in 18″ lengths. Coat all hinge grooves and each piece of cable thoroughly with powdered graphite. Align the hinge members of Assemblies A and B and

ISOMETRIC VIEW

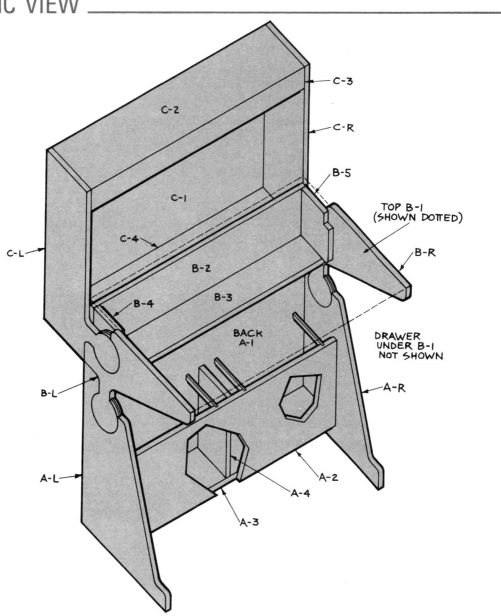

insert a piece of cable in each groove end. Slide hinges together and work mating parts back and forth to move cable through the grooves. Trim the cable ends to length for insertion in the holes. Clean graphite off ends and epoxy them into their holes. Repeat this process with Assembly C. Install panels *B-2*, *B-4*, *B-5* and *B-1*.

Finally, install the drawer. Note: If the alternate drawer slide shown in the detail is used, you must alter drawer dimensions accordingly.

You may wish to weight the pedestal to prevent tipping when the desk is fully opened and accessorized—unless you prefer to anchor the feet to your floor.

HINGE DETAIL

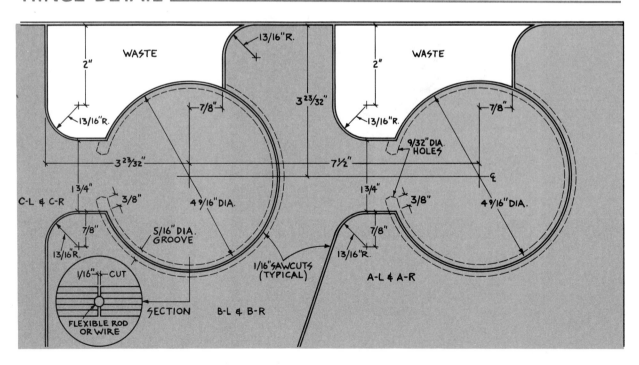

DRAWER

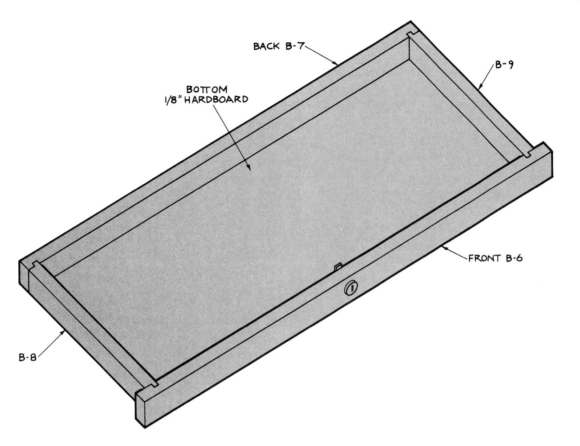

BACK B-7

B-9

BOTTOM
1/8" HARDBOARD

FRONT B-6

B-8

DRAWER DETAILS

Left Side Plan

Drawer Cross Section

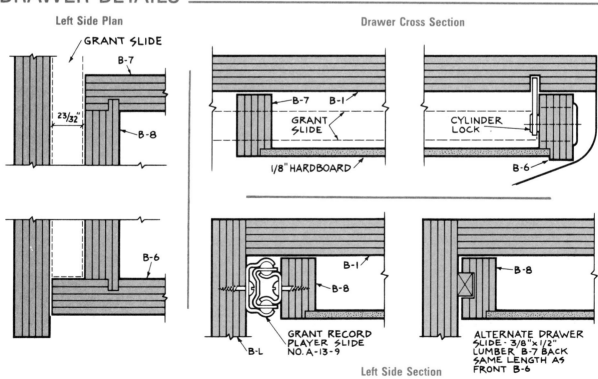

GRANT SLIDE

B-7

23/32"

B-8

B-6

B-7   B-1

GRANT
SLIDE

1/8" HARDBOARD

CYLINDER
LOCK

B-6

B-1

B-8

GRANT RECORD
PLAYER SLIDE
NO. A-13-9

B-L

B-8

ALTERNATE DRAWER
SLIDE - 3/8"x1/2"
LUMBER B-7 BACK
SAME LENGTH AS
FRONT B-6

Left Side Section

DUAL-LEVEL DESIGN CENTER

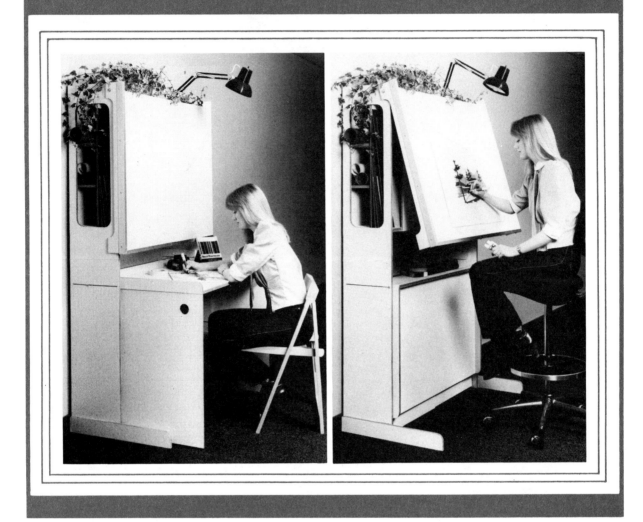

This innovative unit offers two work levels. For the desk height (photo at left), you can ue a folding chair. To work at the adjustable drawing board (right), you'll need a higher perch. When not in use, the unit closes flat (turn page for photo), yet still provides storage in five compartments designed for drawing equipment, rolled plans, and art board.

The lower work surface is also adjustable—to fit the study or hobby needs of the moment. When you really want to spread out, pull this surface all the way out and support it on the cabinet's opened doors. For use as a normal desk top, use it only partially extended.

With proper surfacing—such as slate board or cork—the draw-

Ivan Pavincic
East Rutherford, NJ

PANEL LAYOUTS

$\frac{5}{8}'' \times 4' \times 8'$

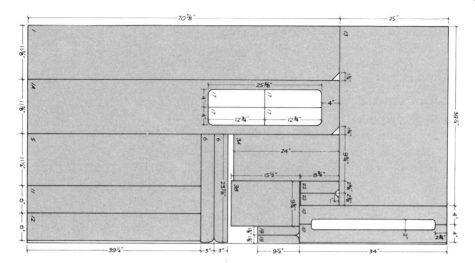

$\frac{5}{8}'' \times 4' \times 8'$

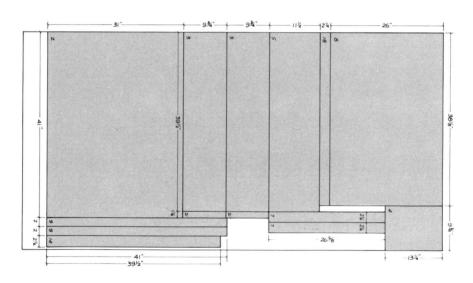

$\frac{3}{8}'' \times 4' \times 8'$

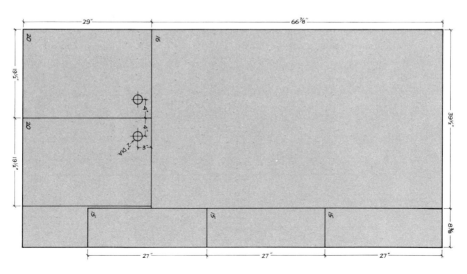

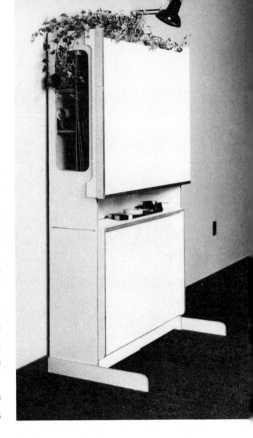

ing board above this desk could double as a demonstration or display surface. The angle of this panel is set by a pair of metal lid supports with knurled-knob tighteners that will lock the board in any desired position.

Parts for the project are cut from three full-size (4' × 8') sheets of plywood—two ⅝″ thick, one ⅜″ thick. When laying out the panels, be sure to number all parts so you can easily identify them during assembly. The designer was so determined to avoid waste that he even utilized the cutout that gives access to the rear storage compartments of the top section. You'll note on the panel layout that this cutout scrap is cut into four equal parts numbered 17. These are simply tucked into the rear corners of the bottom compartment to help support the shelves.

Begin by assembling the two work-surface trays. Each attaches to the uprights with a pivot bolt, but in quite different ways. The top corner of the drawing-board unit has a stationary pivot formed by a lag screw with a washer under its head, driven through the flange from either side.

The flange on the lower surface also has a screw driven into it, but this flange slips *inside* the uprights and the screw slides

▪ MATERIALS LIST ▪

Quantity	Description
1 4 x 8 panel	⅜″ plywood, MDO or A-B Interior
2 4 x 8 panels	⅝″ plywood of similar grade
2	¼″ dia. (shank) x 2″ lag screws w/2 washers each
2	¼″ dia. (shank) x 1½″ roundhead screws w/1 washer each
4	1½″ x 2″ butt hinges w/screws (for cabinet doors)
2	Magnetic catches w/screws (for cabinet doors)
2	Lid supports w/screws (for drawing board)
1	1″ x 2″ x 34″ aluminum angle
1	1″ x 2″ x 40″ aluminum angle
As required	Finishing nails, white or urea resin glue, wood dough or synthetic filler, fine sandpaper, top quality finish

in slots that let you position the desk top various ways and—
when you're through with it—lets you pull it all the way forward
so it will drop and hang in front of the cabinet doors that swing
out to support it when it's fully extended.

We built our prototype of MDO so it would take a smooth paint
coat on its surfaces. We left the edges natural, as a design feature,
merely filling all edge voids, sanding all sharp corners slightly
round and finishing off with polyurethane varnish. You'll want
to paint most parts before assembly, since it would be virtually
impossible to reach inside the many cubbyholes afterward. You'll
find that assembly is much easier than it looks. The best procedure is to leave all edges unfished, so you can glue those that
need it. Then, applying the natural edge treatment is a simple
matter, afterward, since you'll be treating only those that remain
exposed.

FRONT & SIDE VIEWS

Front

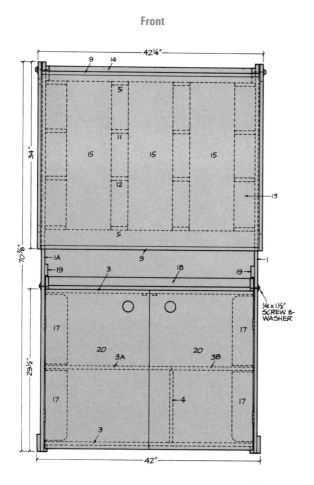

Side

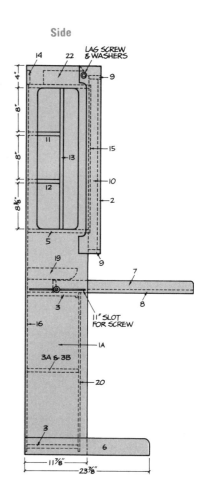

ISOMETRIC VIEW

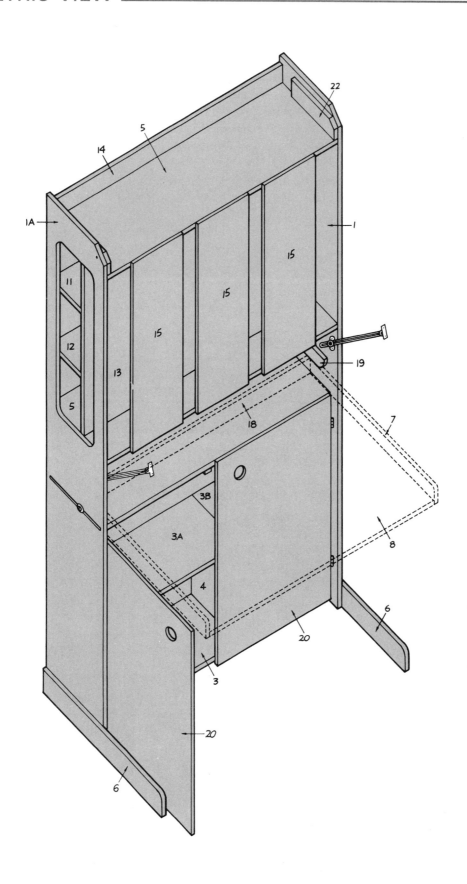

TOP BOLT DETAIL ====================================

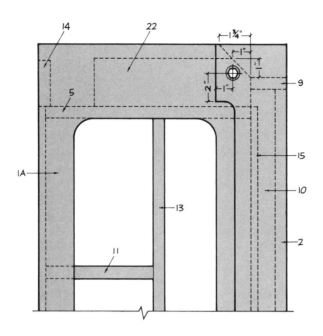

DRAWING BOARD ======================================

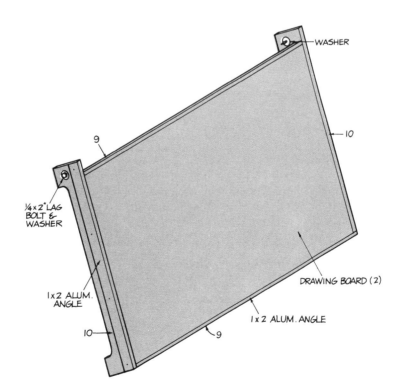

WALL SYSTEM WITH DESK

Divide a room or accent a wall with this modular shelf system that includes a drop-leaf desk. Two sheets of ¾″ MDO plywood will give you the pair of 30″-wide modules shown. Another sheet would add a unit like the one on the left, if you want to go for three modules.

The assembled units are sturdy enough to stand alone, so do not require wall support. Note that there's a good mix between open display shelves and enclosed storage—not only for the desk box but in a double-door compartment to store record albums.

The two-module unit shown here stands over 5′ across and 66″ tall. We cut all parts from MDO panels, painting surfaces white but leaving all exposed edges natural. The stripes of the

Paul Fischer
Apple Valley, MN

124

plies (with voids puttied, edges sanded smooth and coated with oil or with several coats of lacquer) add a sophisticated design accent.

Finger holes are used to eliminate any need for pulls on the cabinet doors and desk lid. The arched tops on the three uprights were cut with a sabre saw, after the U-shaped units were assem-

PANEL LAYOUTS

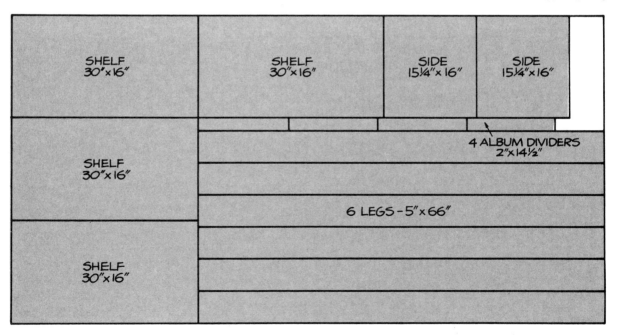

bled with dowels. The corners you trim off the outside (on a 4″ radius) move cleverly inside the assembly to form the inner curve without further cutting.

Shelves and cabinets can be left adjustable (and removable for easy transport to a new location). Each shelf rests on four L-shaped metal supports—the type with a back stud that seats snugly in a hole. Buy this hardware first and drill the holes to match. Stack the uprights and drill all holes through three at once, so they'll align. The holes are on 2″ centers along the centerlines of the 5″ legs. The boxed shelf units (desk and cabinets) are bolted between the leg supports, using machine screws of the proper diameter and length—four to an end panel. You mark the sides for drilling by using the drilled uprights as a template.

Doors are held closed with magnetic catches, as shown in the main assembly diagram (turn page). You'll also need a pair of drop-leaf lid supports. Note that four planks are nailed upright between top and bottom of the cabinet to serve as dividers. We set the backs of our cabinets into rabbets and mitered the corners as shown in a detail, but this is optional.

◢ MATERIALS LIST ▰▰▰▰▰▰

Quantity	Description
2 4 x 8 panels	¾″ plywood, MDO or, if available, A-B Interior
1	Lid support (to support desk front)
4	Magnetic catches (for storage compartment doors)
3 pair	Pin hinges
16	Adjustable shelf supports
8	1¼″ #10 flathead wood screws (for fastening storage compartments to legs)
12	⅜″ dia. x 2″ dowel pins (for joining cross pieces to legs)
As required	Wood dough or synthetic filler, white or urea resin glue, 4d finishing nails, fine sandpaper, top-quality latex paint

NOTES: 1) *Rabbet backs of storage compartments in flush and miter corners as shown in detail drawing.*
2) *Countersink all nails and fill holes—as well as any gaps in exposed plywood edges—with wood dough or synthetic filler. Sand smooth when dry.*
3) *For an attractive effect, accent the edge of the panel with a finishing oil.*

FRONT & SIDE VIEWS

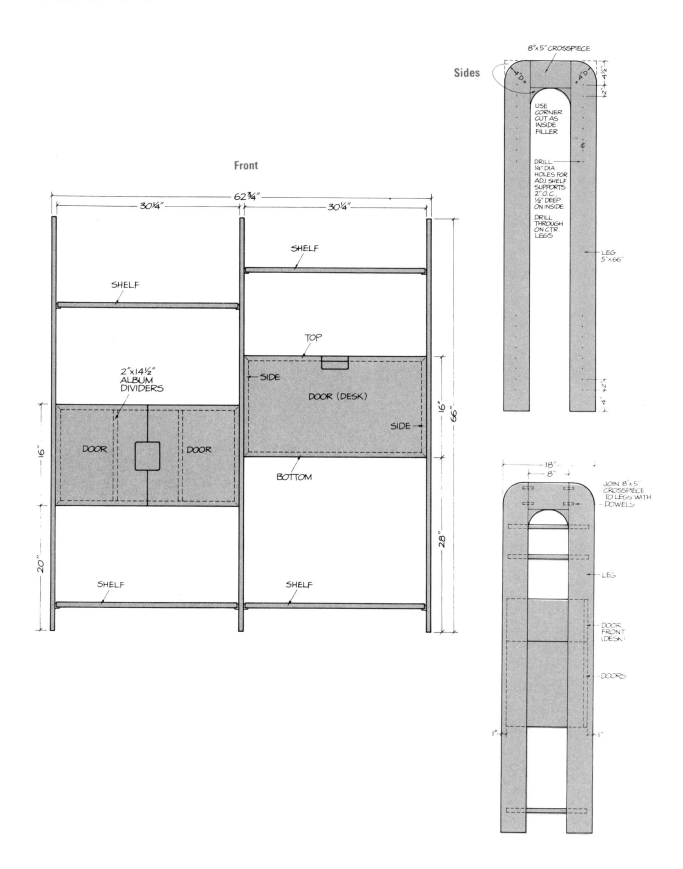

Front

SHELF

SHELF

2"x14½"
ALBUM
DIVIDERS

DOOR

DOOR

SHELF

SHELF

30¼"

62¾"

30¼"

SHELF

TOP

SIDE

DOOR (DESK)

SIDE

BOTTOM

16"

66"

28"

16"

20"

Sides

8"x5" CROSSPIECE

4"D

4"D

4½"

2"

USE
CORNER
CUT AS
INSIDE
FILLER

DRILL
¼" DIA
HOLES FOR
ADJ. SHELF
SUPPORTS
2" O.C.,
½" DEEP
ON INSIDE.

DRILL
THROUGH
ON CTR.
LEGS

LEG
5"x66"

4½"

4"

18"

8"

JOIN 8"x5"
CROSSPIECE
TO LEGS WITH
DOWELS

LEG

DOOR
FRONT
(DESK)

DOORS

1"

1"

ISOMETRIC VIEWS

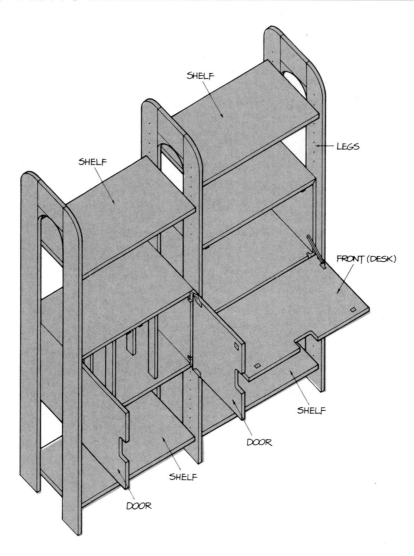

SHELF

SHELF

LEGS

FRONT (DESK)

SHELF

DOOR

SHELF

DOOR

BOX: SIDE & BACK VIEWS

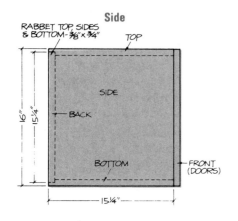

Side

RABBET TOP, SIDES
& BOTTOM - ⅜" x ¾"

TOP

SIDE

BACK

BOTTOM

FRONT
(DOORS)

16"

15¼"

15¼"

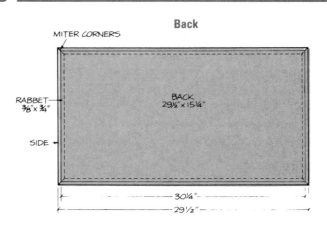

Back

MITER CORNERS

RABBET
⅜" x ¾"

SIDE

BACK
29½" x 15¼"

30¼"

29½"

STUDY CENTER THAT GROWS UP

This clever L-shaped shelf/desk/stool unit for a child's room offers the option of reassembling the components for use by an older student. The stool's legs flip up for greater height when the drop-down desk is repositioned at adult level.

Lay out all parts on two sheets of ¾" plywood. Cut sheets into smaller panels by ripping all-the-way-across lines with a portable circular saw. Other long, straight cuts can be made with this tool, stopping short of corners and finishing cuts with a saber saw. Lay out and cut the nested chair-leg parts carefully. Keep your saw blade centered on the layout lines to equalize saw-kerf waste on all parts. Smooth cut edges with fine sandpaper.

Lay all vertical members face down, side by side; draw hori-

Ron Newbry
Yakima, WA

zontal lines across them to locate the ⅜″-deep dado cuts. Cut these with a router, then drill clearance holes for screws through the grooves, 2″ in from the edges. (If you make screw holes identically spaced, shelves will be interchangeable.) Assemble corner sections (sides and backs) and join them with the top shelf, glued and screwed in place. Fill voids in the cut edges with wood dough and sand smooth when it's dry. Finish all parts before remainder of assembly and attach the desk top to its shelf with a continuous hinge. Insert the shelves into the grooves in the desired arrangement and install screws through predrilled clearance holes and into pilot holes drilled in shelf ends.

Install the 11½″ × 16¾″ desk compartment side panel and glue-screw to shelf B. Drive two screws down through the shelf above the side panel to secure it, gluing *only* if you plan to leave the desk permanently in this location. Install upright front braces and attach desk support chains as shown in exploded views. Install magnetic catches to the upright braces and screw the strike plates in matching positions on the desk top.

To assemble the chair seat, make dado cuts in the seat sides. Glue and screw the back in place, then glue and screw the seat on. Assemble the chair base to desired height, using only screws

MATERIALS LIST

Quantity	Description
2 4 x 8 panels	¾″ plywood, A-B or A-D Interior, A-B or A-C Exterior, or MDO
72 approx.	#8 flathead wood screws, 1¼″ long for shelves, desk compartment and frame, and chair back
15 approx.	#8 flathead wood screws, 1½″ long for chair legs and fastening chair bottom to legs
1	Piano type (continuous) hinge, 34″ long for desk top, plus screws for fastening
2	Magnetic catch and striker sets for desk closure
32″	Small-link chain for two 16″ desk top supports
4	Eye screws for fastening chains
As required	Wood dough for filling any small voids in plywood cut edges; fine sandpaper for smoothing cut edges and cured wood dough
As required	White glue for glue-screw assembly
As required	Finishing materials: primer and paint; antiquing kit; or synthetic satin-finish varnish (e.g., Varathane), with or without stain

if you plan to change it later. Mount the seat on its base.

When changing from one desk location to the other, you'll have to fill unused screw holes with wood dough, sand flush and apply a touch-up finish.

PANEL LAYOUTS

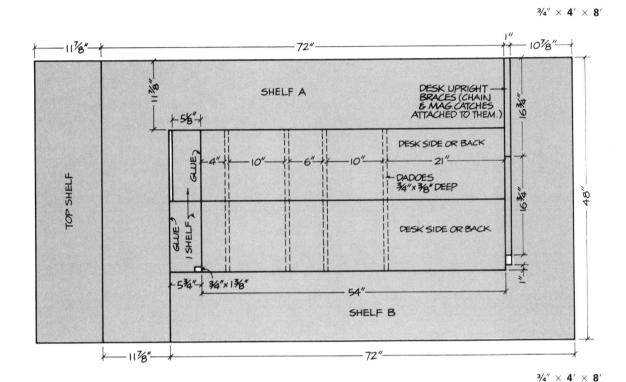

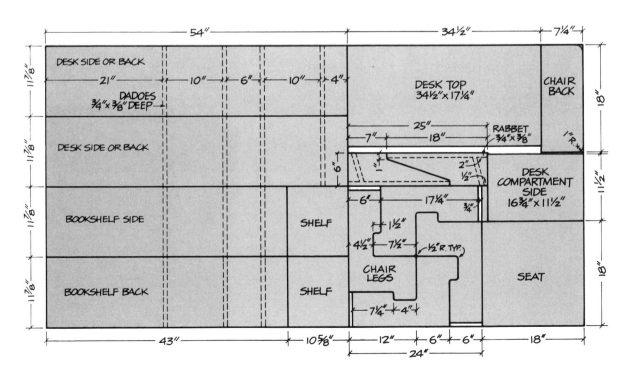

CHILD'S VERSION DESK

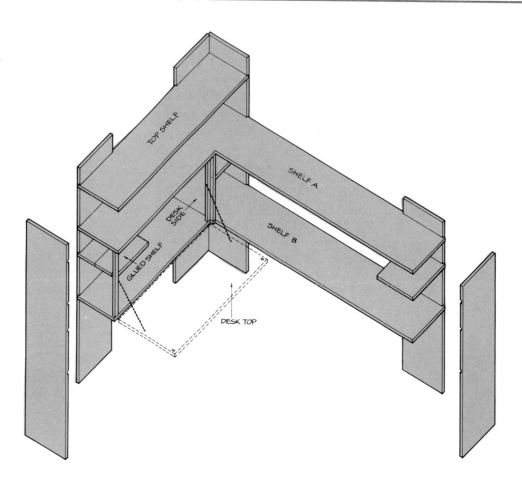

TOP SHELF

SHELF A

DESK SIDE

SHELF B

GLUED SHELF

DESK TOP

CHILD'S VERSION SEAT

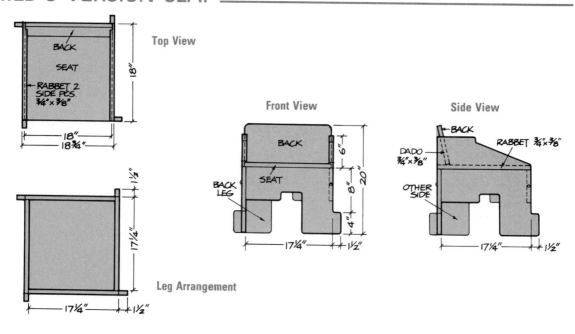

Top View

BACK

SEAT

RABBET 2 SIDE PCS. ¾" x ⅜"

18"

18"
18¾"

1½"

17¼"

17¼"   1½"

Leg Arrangement

Front View

BACK

SEAT

BACK LEG

6"

20"

8"

4"

17¼"   1½"

Side View

BACK

RABBET ¾" x ⅜"

DADO ¾" x ⅜"

OTHER SIDE

17¼"   1½"

TEEN'S VERSION DESK

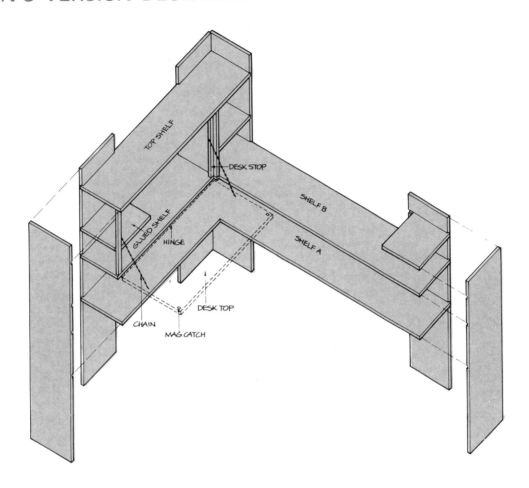

TEEN'S VERSION SEAT

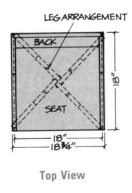

Top View

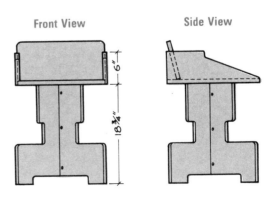

Front View   Side View

CONVEX CUBBYHOLE SHELF

David J. Greene
Philadelphia, PA

Think of it as a slice off an egg-crate ball. As the structure bows out into the room it presents ever deeper compartments for a variety of storage needs. Though each cubbyhole measures about 8½″ square, the shallower ones around the rim display small items that might get lost behind bulkier ones. And because the shelves are convex, you can even store, on the upper shelves, long-necked items that will stick up past the recessed shelf above them. In short, the varying depths of the compartments help organize your storage and display by determining the objects or books each will accept.

Since fit is critical for the rigidity of the unit, it's best to make cardboard templates for each size shelf, and trace around these to do your panel layout. You need four pieces of each size, two of each with slots cut from the flat back and the other pair with slots cut from the curved front edge. Cut only the shelf outlines first; you'll cut the slots later.

Stack same-size shelves and clamp together in sets of four. Fill all edge voids, sand edges smooth and chamfer all sharp corners to prevent chipping.

If you plan to stain the shelves a natural wood-tone, as in the photo, you'll want to apply veneer tape to all the curved edges that will be exposed when the unit is hung. Trim away excess when the glue has set.

Clamp shelf sets back together, making certain all edges align. Mark slots with a square as follows: Mark exact centerline on back edges of all sets; mark midlines of slots 4½" to either side of the centerline. Mark next midlines 9" from first two slots. Mark slot widths, ⅜" on each side of midlines for a total width of ¾". Remove clamps and carry slot lines across panel faces to the front curved edges. Measure along each slot and mark off half its total length. Cut slots in one pair of each set from back edge to mid-point, and slots in the other pair from the front edge. If you don't

ISOMETRIC VIEW

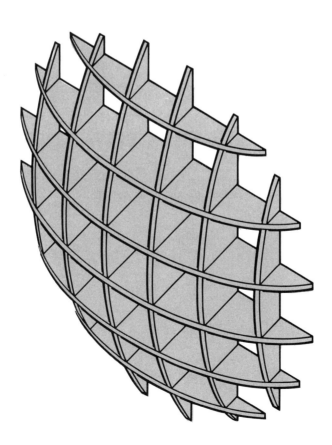

have a router, cut the slots using a fine-tooted saw and a chisel. Finish all parts before slipping them together. For permanent assembly, leave finish off the slot-mark channels and apply glue while slipping parts together.

PANEL LAYOUT

¾″ × **4**′ × **8**′

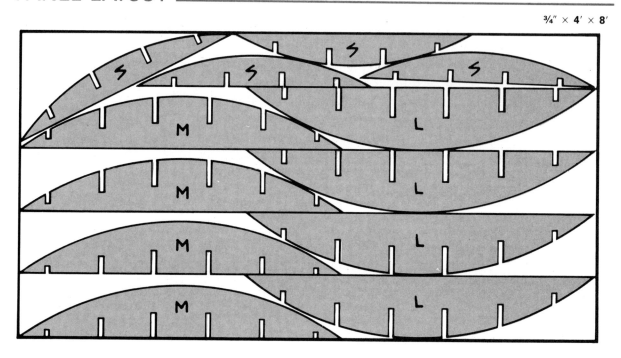

MATERIALS LIST

Quantity	Description
1 4 x 8 panel	¾″ plywood, A-A or A-B Interior or MDO
3	4″ x 5″ shelf brackets and fasteners (see note)
50′ approx.	Wood veneer trim, 1″ wide, light color such as birch
Small can	Contact cement for applying veneer
As required	Wood dough or other filler for filling voids in plywood cut edges
As required	Fine sandpaper for smoothing plywood cut edges; finishing materials

Note: If fastening to drywall, Mollybolts or toggle bolts will be necessary.

PATTERNS FOR TEMPLATES

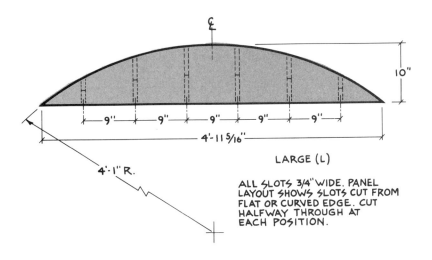

LARGE (L)

ALL SLOTS 3/4" WIDE. PANEL
LAYOUT SHOWS SLOTS CUT FROM
FLAT OR CURVED EDGE. CUT
HALFWAY THROUGH AT
EACH POSITION.

10"
9" 9" 9" 9" 9"
4'-11 5/16"
4'-1" R.

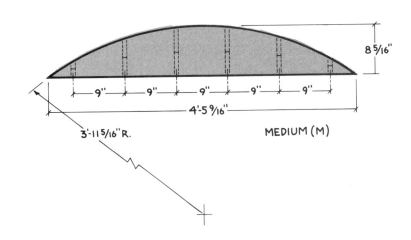

MEDIUM (M)

8 5/16"
9" 9" 9" 9" 9"
4'-5 9/16"
3'-11 5/16" R.

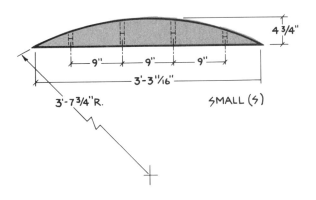

SMALL (S)

4 3/4"
9" 9" 9"
3'-3 11/16"
3'-7 3/4" R.

FRONT & SIDE VIEWS

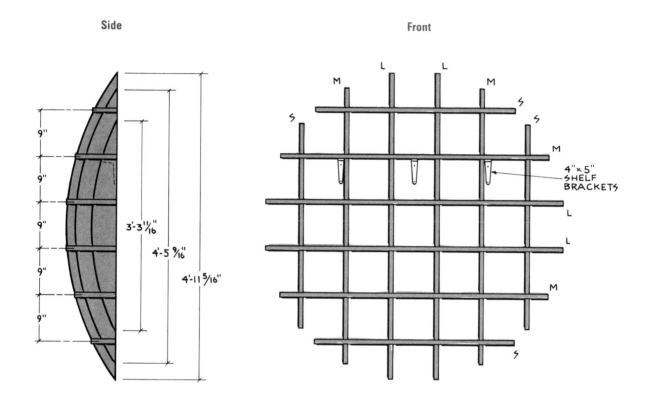

Side

Front

9"

9"

9"

9"

9"

3'-3 11/16"

4'-5 9/16"

4'-11 5/16"

L   L

M   M

S   S

S

M

4" × 5"
SHELF
BRACKETS

L

L

M

S

SLOT DETAIL

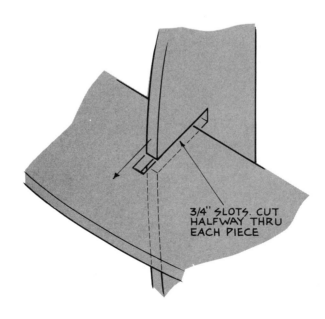

3/4" SLOTS. CUT
HALFWAY THRU
EACH PIECE

SWING-WING LIQUOR CABINET

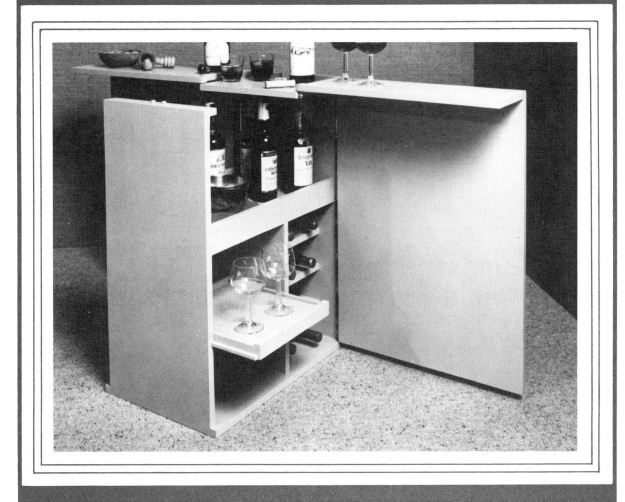

Maybe you're a host who doesn't like to flaunt liquor and wine, especially when guests happen to be non-drinkers. For such occasions, this shrewdly designed chest closes up into a totally noncommittal box that's neat enough to take its place in any living room or den (see photo next page). It can be a handsome piece of furniture if you cut at least the outside panels from cabinet-grade plywood, or veneer the five exposed surfaces, enameling the edges in a contrasting color. (The designer/winner built his original prototype from a single sheet of MDO and painted the whole unit a warm gray.)

This unit is more than a storage cabinet—it's a caddy that keeps an ample supply of different beverages within easy reach. The

Norbert G. Marklin
St. Louis, MO

liquor stands upright, the wine lies on its side, keeping the corks properly moist. There's also a slide-out shelf for glassware and space enough to tuck in corkscrews and ice tongs. When the unit spreads its wings, pivoting them right and left, it forms a good-sized service bar.

The designer holds his cabinet open or closed with simple but handsome pegs; he turned all four from a ½"-dia. birch dowel centered in a hobby lathe, then sawed them apart. Similar pegs, passing down through screw-eyes, secure the wings in their open position. The wings hang from continuous (piano) hinges.

Lay out all plywood pieces as shown. After cutting, be sure to relieve the inside corners of the door-top tabs, as shown in the detail. With a router (or on a table saw) cut the rabbets across the inside faces of the front and the two tabbed tops. Drill holes in the "front top" ledge, in the tab cutout of the upper edge of the back, and through the corners of the top panels. These latter holes can be counterbored, as shown, to sink part of the peg head, but this is optional.

Next, assemble the sliding glass tray. Note that the tray can slide on 1 × 1 cleats (as shown in the drawings) or can have its bottom panel project at each side into grooves cut across the inner faces of its compartment sides. If you choose the latter, make this tray bottom of ¼" hardboard—preferaby the type with both faces smooth. Either way, you'll want to provide a stop block that will

PANEL LAYOUT

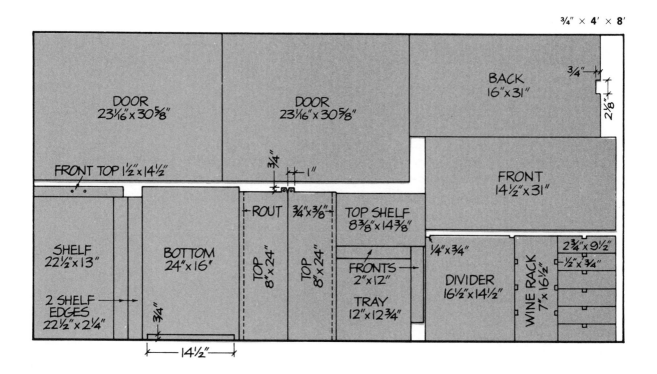

MATERIALS LIST

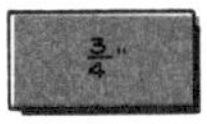

Quantity	Description
1 4 x 8 panel	¾" plywood, MDO or, if available, A-B Interior
30 lin. in.	1 x 1 lumber for sliding tray guides and stops
6½ lin. in.	½" dia. hardwood dowel
2	30⅝" x ¾" piano hinges with screws
2	2" x 2" right angle metal shelf supports with screws
2	1" x ⅜" inside dia. screw eyes
As required	4d finishing nails, white or urea resin glue, wood dough or synthetic filler, fine sandpaper, top-quality enamel paint with companion primer

FRONT & SIDE VIEWS

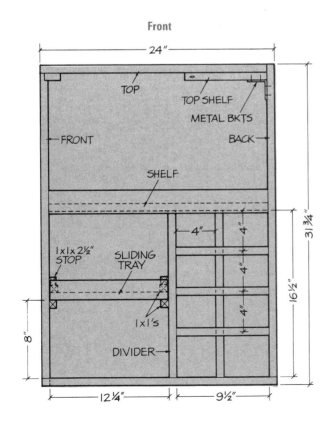

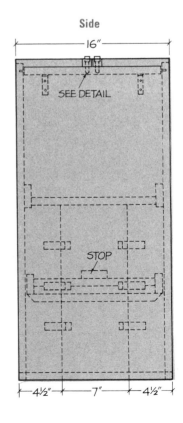

catch the tray's back flange to prevent it from pulling free of the cabinet.

Now assemble all cabinet parts, glue-nailing all joints, starting with the front (which sets into the notched edge of the bottom), the back and the shelf. Once the sliding shelf is in place, do final fitting and assembly of the wine rack. Hinge the two door assemblies in place, install the screweyes to mate up with the peg holes, then close each door assembly to see where the slit must be chiseled on the inner face to accept the projecting screweyes.

TOP VIEW

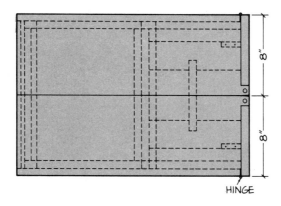

BACK PIN DETAILS

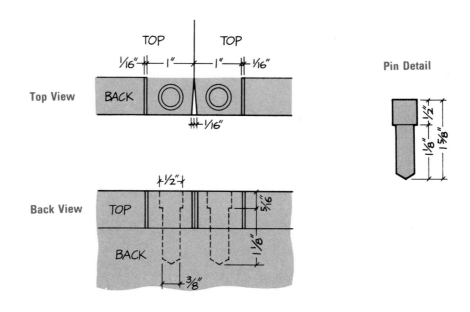

EXPLODED VIEW

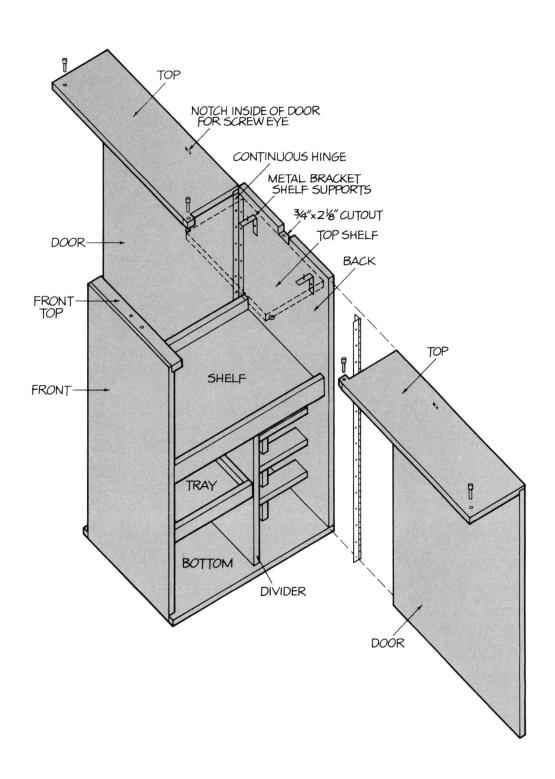

TOP

NOTCH INSIDE OF DOOR
FOR SCREW EYE

CONTINUOUS HINGE

METAL BRACKET
SHELF SUPPORTS

3/4" x 2⅛" CUTOUT

DOOR

TOP SHELF

BACK

FRONT
TOP

FRONT

TOP

SHELF

TRAY

DOOR

BOTTOM

DIVIDER

WINE CENTER

This wine center is as handy on a shaded patio as in a dining room. The "egg-crate" rack stores two cases of wine and has bins for bottles of unusual shape. The slotted shelves keep a dozen or so stem glasses ready for service. You'll note that the photo shows six slots per shelf, while the plans call for seven; it depends on the diameters of your glasses' bowls and bases. One early caution: Be sure to paint the rack parts prior to assembly. It's a tedious job afterward.

With an overall size of about 1' × 2' × 3½', the rack can be free-standing, mounted as a wall unit, or used as a portable bar. Its sturdy glue-nail construction makes it durable for years of service. Lay out all parts on your ½" plywood panel. For ease in

Patrick Nyby
Woodland Hills, CA

cutting the bottle-rack slots, clamp cut partitions together in matching stacks (four 6″ × 16″ parts with three slots each, two 6″ × 17″ parts with one slot each, etc.). Mark the exact centerline of slots on the panel edges and cut (preferably with a back saw) an accurate kerf ¼″ to each side. Unclamp the stacks and complete the layouts of all slots, cutting them with a saw and chisel—or (more quickly) with a router.

To cut the glass-hanger slots, clamp the two matched parts together with their front edges offset one inch and, after laying out the slots, drill a ⅝″ hole through the stack, at the inner end of each slot. Cut two kerfs into the edges of each hole to form the rounded slots. Assemble the bottle unit to check all pieces for fit, making any necessary adjustments. Disassemble for finishing, then apply glue to the slots. Lay the side and back members side-by-side and draw lines across them to mark shelf and glass rack locations. Using glue and nails, join back, sides, bottom and top as shown in the exploded view.

Glue-nail the glass hanger rack and shelf in place and install the mirror tiles (two 12″ squares). Countersink all exposed nail heads, fill and sand flush. Finish the case as desired. When dry, set the prefinished bottle rack in place, centered in the cavity; secure it by toe-nailing through pre-drilled holes. Install the 1 × 2 face trim with glue and nails. We enameled our prototype a forest green, but since the bottle rack is finished separately, you may prefer a two-tone effect.

PANEL LAYOUT

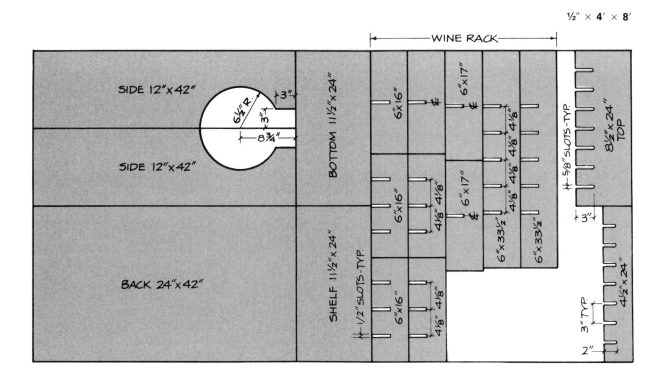

MATERIALS LIST

Quantity	Description
1 4 x 8 panel	½" plywood, A-B or A-D Interior, A-B or A-C Exterior, or MDO
4 pieces	1 x 2 lumber (¾" x 1½"), 25" long for face trim, mitered at corners
2 (optional)	12" x 12" mirror tiles
20' approx.	½" wide wood veneer trim for bottle-rack edges, unless painted
½ lb approx.	4d finishing nails for glue-nail assembly
As required	White glue (approx. 8 oz can) for glue-nail assembly
As required	Fine sandpaper and wood dough (if desired) for plywood cut edges; finishing materials

SIDE & FRONT VIEWS

Front

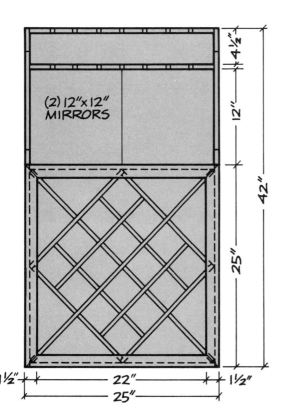

Side

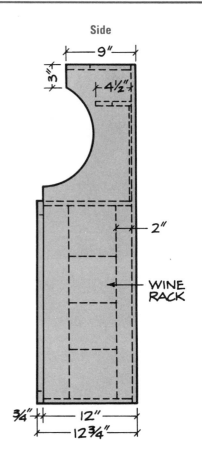

EXPLODED VIEW

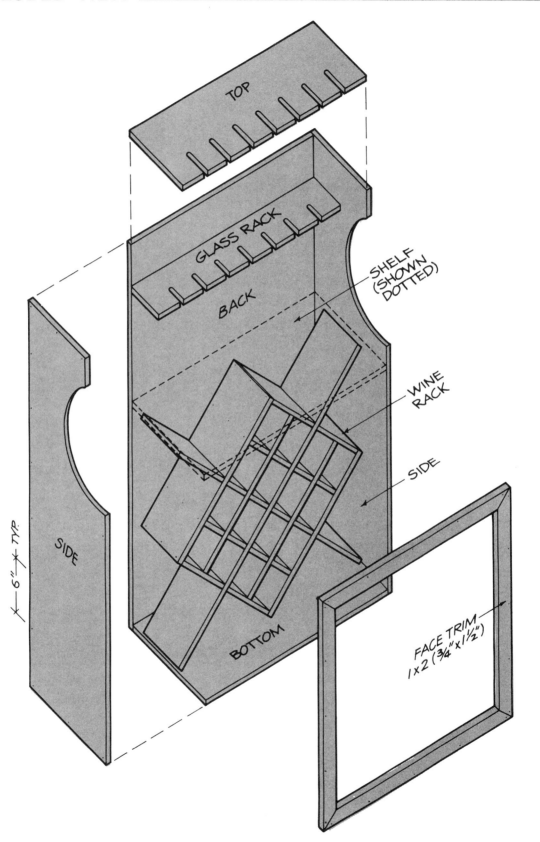

TOP

GLASS RACK

BACK

SHELF
(SHOWN
DOTTED)

WINE
RACK

SIDE

SIDE

6" TYP.

BOTTOM

FACE TRIM
1x2 (¾"x1½")

RECORD AND CASSETTE RACK

All parts for this rack for albums, cassettes and cartridges are cut from a half sheet of ¾″ plywood. They slot together egg-crate fashion so the entire assembly can be taken apart for portability. It's an ideal organizer to keep near stereo equipment.

Construction couldn't be simpler: Note on the panel layout that one line is labeled "First Cut". That separates your half sheet into two basic units. A second cut across the larger of these units gives you three sections. Then it's an easy matter of slicing two of the sections into three parts each.

Now comes the only hard part—cutting the slots. But this goes quicker than you'd think because the three *A* panels are identical, and if you're careful—and have the proper tools—you can cut all these slots simultaneously. Just stack the three panels and

Lauren Drier
Parma Heights, OH

tape them firmly together (wide masking tape is best). Now, on your drill press, bore ¾" holes through the stack at the bottom of each of the slot layouts. Carry the stack to your bandsaw and cut into both sides of each hole along the layout lines. You can square the bottom corners after you've separated the panels again. Follow the same procedure for the C panels. (The B and D panels should be slotted individually.)

The finish you'll apply will determine the exact slot width. We built our prototype of MDO and enameled all parts in two contrasting colors. (All the upright A panels were tan, the other panels a milk-chocolate.) Painting increases the thickness enough so the assembly would bind if the slots were not slightly larger than your plywood thickness. If you choose to apply a simple natural finish or a stain, you needn't provide this extra thickness. But give special attention to filling any edge voids and sanding all edges smooth and slightly round before applying any finish.

Once the assembly is complete, you'll have two bins for record albums (pop and classical?) plus four compartments for cassettes on one side and two deeper compartments for cartridges on the other. And the whole thing sits pertly up off the floor or countertop to ride above any accumulated dust.

PANEL LAYOUT

¾" × **4'** × **4'**

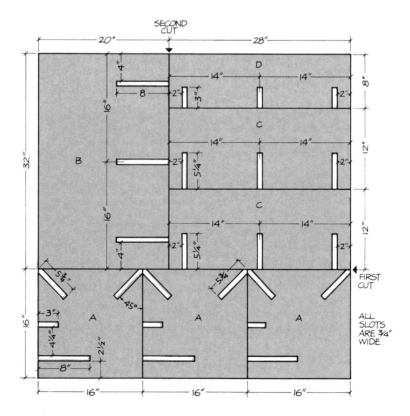

MATERIALS LIST

Quantity	Description
1 4 x 4 panel	¾″ plywood, MDO or, if available, A-B Interior
As required	Wood dough or synthetic filler, fine sandpaper, top-quality latex paint

TOP, FRONT & END VIEWS

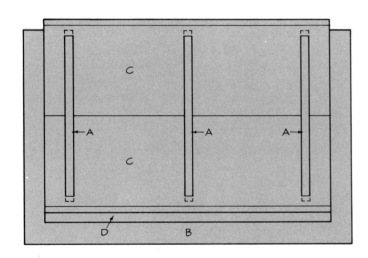

Top

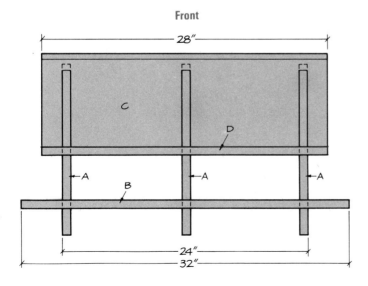

Front

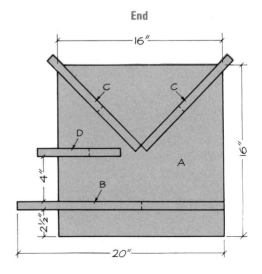

End

VISE TABLE AND SAWHORSE

Make yourself this pair of workshop accessories and you'll pave the path to many other projects in this book. All parts are cut from a single sheet of plywood and they're so shrewdly nested there's no waste worth mentioning. These parts assemble into two portable, compact, and stowable units: a split-top vise table and a matching-height sawhorse that's ideal for supporting long stock.

The table, closed, is a solid level surface, 1½" thick, for workshop jobs ranging from planing to wood-carving. As a vise, it boasts a 4" capacity. The vise mechanism is quite simple to assemble from threaded rod, angle iron, nuts, and bolts.

After you've laid out your parts as shown, cut off the table-top section, then cut it in half and set these two sections aside while

Ron Newbry
Yakima, WA

you cut the remaining parts. To cut slots, clamp or tack the saw-horse legs and ends into pairs and cut slots simultaneously—preferably on a bandsaw. After sanding all cut edges, slide the sawhorse ends into the leg slots to check the fit. When satisfied, fasten ends to top with hinges.

Laminate the two table-top sections together, gluing the mating surfaces and clamping until dry. Now cut this laminated panel lengthwise into two 12″ sections. Assemble the runner channels from the side and bottom plates and fasten to the stationary section of the table top, as shown in a detail. Be careful to keep these channels parallel so the slides won't bind. Use glue and flathead screws: 2″ screws where channels fasten to the top, 1¼″ screws on the other half. Attach the spacer plate using glue and 1¼″ screws.

Place the runner slides in the channels and lay the movable table section in position on top of the slides. Mark the position of the slides on the underside and then, using 1¼″ screws (plus glue), fasten the slides to the movable section as shown in the exploded view. This assures the slides will run true.

Hinge the legs to the stationary top assembly, as shown in the bottom and front views. Note that the spacer plate on one side lets the legs fold flat. Also install the hinges that fasten the cross braces to the legs (see end and front views). When these braces overlap, drill a 5⁄16″-dia. hole through them. Insert carriage bolts and wing nuts—removable for storage.

PANEL LAYOUT

¾″ × 4′ × 8′

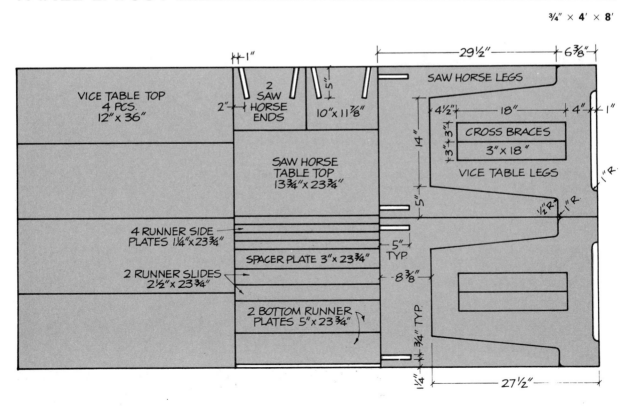

MATERIALS LIST

¾"

Quantity	Description
1 4 x 8 panel	¾" plywood, A-B or A-D Interior, A-B or A-C Exterior, or MDO
12	Strap hinges, 1½" with screws
4 pieces	Angle iron, 1½" x 2"
2 pieces	Threaded rod, ⅜" dia. x 18" long
4	Nuts, to fit ⅜" dia. threaded rod
2	Bolts, ¼" dia. x 5" long
2	Lock nuts, ⅜" dia. x ½" deep
20 approx.	#8 flathead wood screws, 1¼" long
8 approx.	#8 flathead wood screws, 2" long
8	#8 roundhead wood screws, 1¼" long
2	Carriage bolts, ¼" dia. x 2" long, with wing nuts
4	Compression pins, ⅛" dia. x ½" long
As required	Wood dough for filling any small voids in plywood cut edges; fine sandpaper for smoothing cut edges and cured wood dough
As required	White glue for glue-screw assembly
As required	Finishing materials

For the vise mechanism, drill a 3/16" hole through one facet of two lock nuts; tap with ¼" threads. Lock these at one end of each threaded rod; turn a bent 5" bolt into the side holes to form handles. Drill two 3/16 holes in one flange of each angle iron to take 1¼" roundhead mounting screws. Center a 5/16" hole in the other flanges of the irons to be mounted on the movable table section and tap to match the threaded rod. Center a 5/16" hole in the other flanges of the angles to be mounted on the stationary table. Screw these to the bottom faces of the table sections, flush against the inside edges of the runner channels; align the rod holes carefully. Install the two rods as shown, then drill ⅛" holes through the four retaining nuts and rods. Insert compression pins to lock the nuts in that position. Apply a good spar varnish, but not to the runner channels or slides. Rub these with paraffin.

TABLE: EXPLODED VIEW

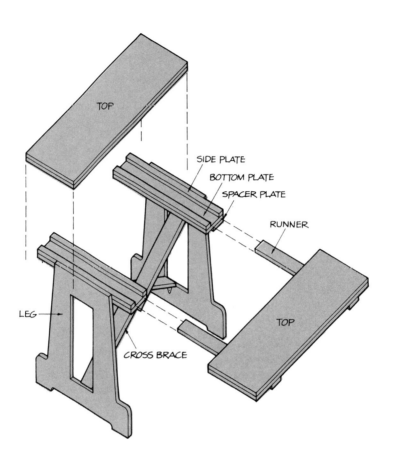

TOP

SIDE PLATE

BOTTOM PLATE

SPACER PLATE

RUNNER

LEG

TOP

CROSS BRACE

TABLE: SIDE & FRONT VIEWS

Side

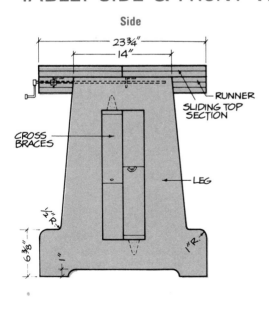

23¾"

14"

RUNNER

SLIDING TOP SECTION

CROSS BRACES

LEG

½"R.

6⅜"

1"

1"R.

Front

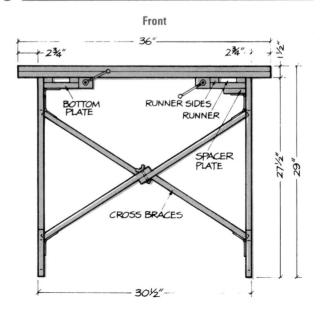

36"

2¾"

2¾"

½"

BOTTOM PLATE

RUNNER SIDES

RUNNER

SPACER PLATE

CROSS BRACES

27½"

29"

30½"

TABLE: TOP VIEW

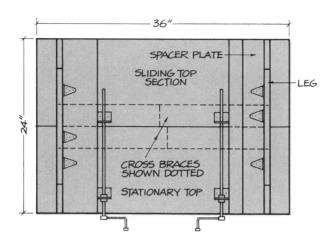

36"

24"

SPACER PLATE

SLIDING TOP SECTION

LEG

CROSS BRACES SHOWN DOTTED

STATIONARY TOP

TABLE: VISE ASSEMBLY

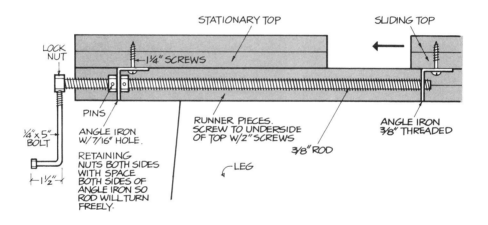

STATIONARY TOP

SLIDING TOP

LOCK NUT

1¼" SCREWS

PINS

¼" x 5" BOLT

ANGLE IRON W/ 7/16" HOLE.

RETAINING NUTS BOTH SIDES WITH SPACE BOTH SIDES OF ANGLE IRON SO ROD WILL TURN FREELY.

1½"

RUNNER PIECES. SCREW TO UNDERSIDE OF TOP W/2" SCREWS

3/8" ROD

LEG

ANGLE IRON 3/8" THREADED

TABLE: FRONT VIEW DETAIL

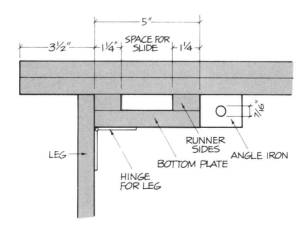

5"

SPACE FOR SLIDE

3½"

1¼"

1¼"

7/16"

LEG

HINGE FOR LEG

BOTTOM PLATE

RUNNER SIDES

ANGLE IRON

SAWHORSE: EXPLODED VIEW

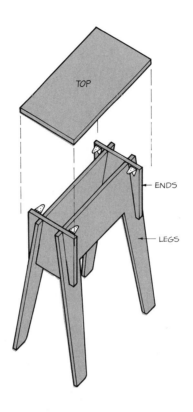

TOP

ENDS

LEGS

SAWHORSE: FRONT, SIDE, & BOTTOM VIEWS

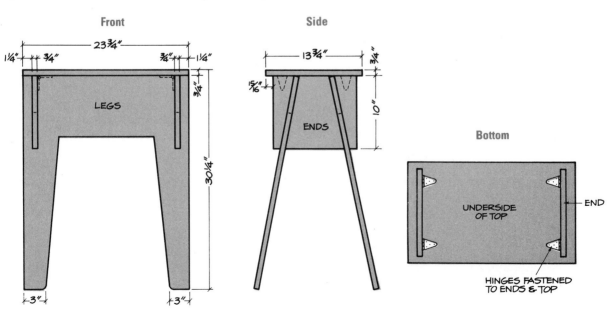

Front

23¾"

1¼" ¾" ¾" 1¼"

¾"

LEGS

30¼"

3" 3"

Side

13¾"

¾"

15/16"

10"

ENDS

Bottom

UNDERSIDE
OF TOP

END

HINGES FASTENED
TO ENDS & TOP

FOLD-DOWN WORKBENCH

Is space too tight in your home or garage for a bulky workbench? This one was designed for the side wall of a garage: Back out the car and the bench unhooks and drops for use, complete with a full-width, built-in vise. It's an equally handy unit anywhere that floorspace is at a premium. All pieces for the bench, the vise and the U-shaped wall bracket were cut from a single sheet, with only sawdust remaining.

With the proper fasteners (lag screws, molly or toggle bolts), the wall plate and bracket can be attached to any type of wall. For masonry, you'd need lead anchors, but for other materials your choice depends on whether the wall is solid or hollow.

The double lamination of the plywood top, vise and leg-brace

Martin Gierke
Burnsville, MN

members makes for great strength and durability—plus the sturdiness that's essential in any work surface.

The hardware itemized in the materials list is readily available. The ⅜″ steel rod is cut into four lengths and bent into L-shapes to serve as handles for the vise screws and as lock bolts for the wall bracket. Note that, to keep the latter within easy reach, the bracket is mounted well below where the end of the bench comes in its flipped-up position. For vise handles, the L-shaped rods are inserted through holes drilled near the ends of each threaded rod, and soldered or welded in place. In fact, it's best to also run them through a hex nut turned flush onto this end of the rod (drill through both nut and rod at the same time). The opposite ends of the threaded rods must also be drilled through, since they pass through a drag bar sandwiched between washers; the rods are locked in this position by cotter pins inserted through the holes drilled just outside the two washers on each rod. Drill each threaded rod a fourth time for a third cotter pin that locks other washers against the inside of the vise's moving jaw.

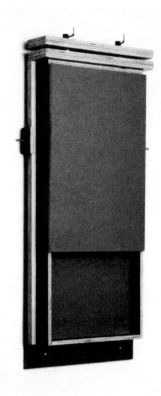

Note also that the rod turns through a long nut embedded in the piller plants behind the stationary jaw.

Because of the weight of the vise parts in our prototype, we found that the hinged leg sits firmly on the floor and requires no folding brace to keep it in position—so this is optional and does not appear in the materials list. Note that all plywood parts are numbered in the list to correspond to the numbers on the cutting diagram.

PANEL LAYOUT

¾″ × 4′ × 8′

1 TOP 48″ x 22½″		3 LEG 31½″x 21″		4	5	6	7
		8 / 9	10	11	12	13	
		14		15			
2 TOP 48″ x 22½″		16		17			
		18	19		24		
		20	21		25		
		22	23		26		
27		28					

MATERIALS LIST

¾"

Quantity	Description
1 4 x 8 panel	¾" plywood, MDO or A-B Interior, A-C Exterior, or B-B Interior or Exterior
Item No.	*(All measurements below in inches)*
1, 2	Top 48 x 22½
3	Front "Leg" 31½ x 21
4, 5	Leg Brace 19½ x 3
6	Wall Plate 24 x 6
7	Wall Bracket 24 x 4½
8, 9	Drag-Bar Support 24 x 1½
10, 11	Nut Back-up 3¾ x 3
12, 13	Bracket Side 4½ x 3
14-17	Vise 24 x 6
18, 19	Drag Bar 19½ x 3
20-22	Front Brace 19½ x 3
23	Rear Brace 19½ x 3
24-26	Center Brace 9 x 3
27, 28	Side Brace 48 x 3

OTHER MATERIALS

Approx. 2 lin. ft	⅜" steel rod for vise handles (2) and wall brackets (2) (Cut and bend as shown; drill threaded rod and solder to vise handles)
2	¾" x 24" threaded rods
2	¾" x 1⅛" nuts (for ends of threaded rods)
2	¾" x 2¼" nuts (for backup)
6	Cotter pins
4	3" x 3" hinges with screws
8	¾" washers
2	1½" x ¾" x 48" optional wood strips (to protect top edges)
As required	Finishing nails, white or urea resin glue, wood dough or synthetic filler, sandpaper, top quality finish, if desired

EXPLODED VIEW

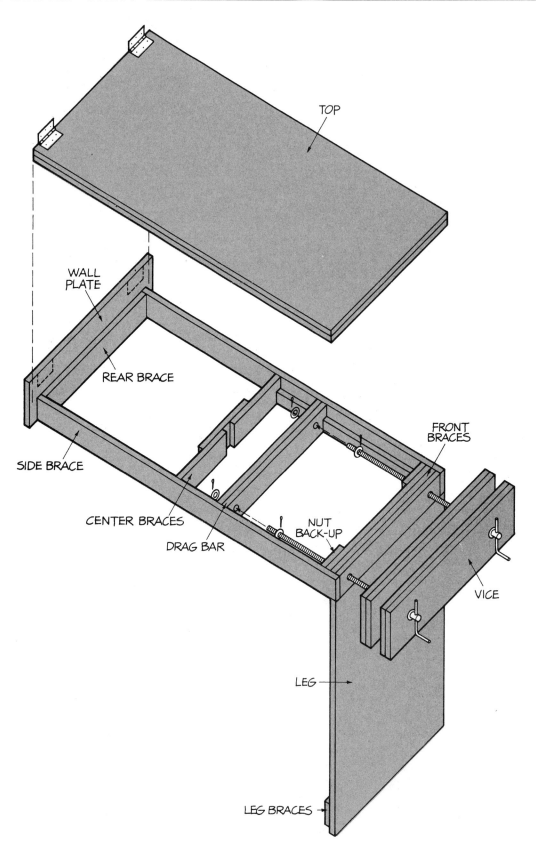

TOP

WALL
PLATE

REAR BRACE

SIDE BRACE

FRONT
BRACES

CENTER BRACES

DRAG BAR

NUT
BACK-UP

VICE

LEG

LEG BRACES

FRONT & SIDE VIEWS

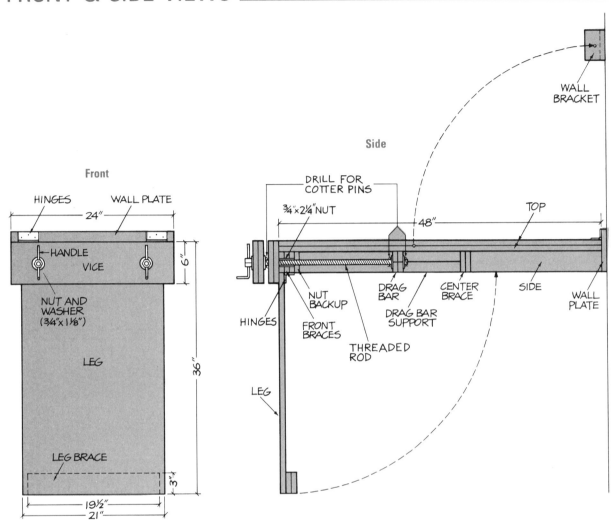

Front

HINGES WALL PLATE
24"
HANDLE
VICE
6"
NUT AND WASHER
(3/4" x 1 1/8")
LEG
36"
LEG BRACE
3"
19 1/2"
21"

Side

WALL BRACKET

DRILL FOR COTTER PINS
3/4" x 2 1/4" NUT
48"
TOP
NUT BACKUP
DRAG BAR
CENTER BRACE
SIDE
WALL PLATE
HINGES
FRONT BRACES
DRAG BAR SUPPORT
THREADED ROD
LEG

WALL BRACKET DETAIL

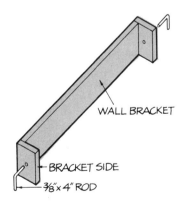

WALL BRACKET
BRACKET SIDE
3/8" x 4" ROD

PAIR OF KNOCKDOWN SAWHORSES

One sawhorse does you little good—you need a good steady pair that won't tip over at that awkward moment when you're struggling to slide a long panel or plank on top of them.

This pair is as cleverly designed a set as I've seen—and all parts cut from a half-sheet of ¾" plywood. The great bonus is that they fold flat: There's nothing bulkier than standard sawhorses squatting in your way when you don't need them.

These are quite decorative, too, so we chose a good Exterior panel (A-C) and after cutting all parts, filled voids in the edges, sanded them smooth and enameled them burnt orange—a nice contrast with faces left natural (either oiled or brushed with spar varnish; if you apply the face finish first, it's easier to wipe off any drips or mistakes of the edge color).

Kevin A. Thorp
Summerville, SC

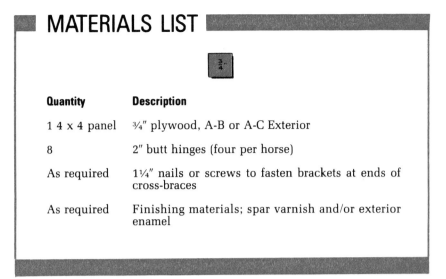

MATERIALS LIST

Quantity	Description
1 4 x 4 panel	¾″ plywood, A-B or A-C Exterior
8	2″ butt hinges (four per horse)
As required	1¼″ nails or screws to fasten brackets at ends of cross-braces
As required	Finishing materials; spar varnish and/or exterior enamel

The pattern shown below is for a single horse. You make the identical layout twice on the 4×4 plywood sheet. After cutting (you'll need a jig-, band- or sabre saw), glue-nail the slotted disk-like parts to the ends of the 3″-wide cross-braces as shown, then hinge the 6″ brace between the leg units.

To set up a horse, just swing the legs at right angles to the center brace and secure them in that position by sliding a foot brace into the slots at both sides.

PANEL LAYOUT & EXPLODED VIEW

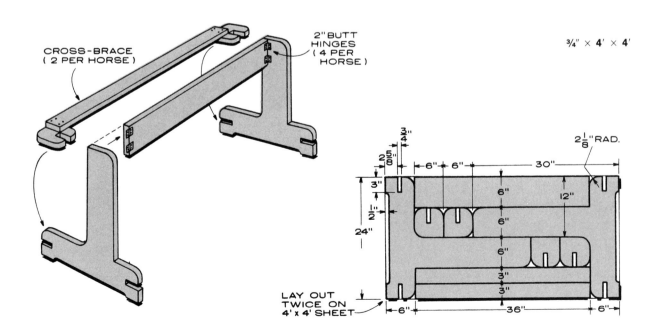

CROSS-BRACE (2 PER HORSE)

2″ BUTT HINGES (4 PER HORSE)

¾″ × 4′ × 4′

LAY OUT TWICE ON 4′ x 4′ SHEET

2⅛″RAD.

SEWING CENTER

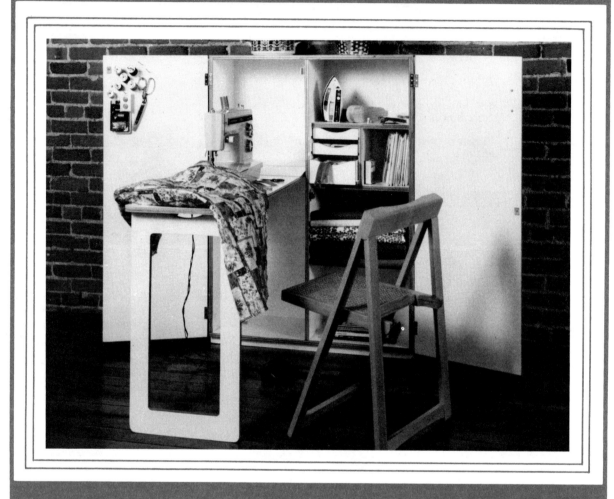

When not in use, this well-planned sewing center folds up into a neat cabinet only 1′ deep and 3′ wide. Yet unfolded it offers a generous work surface and ample storage for materials, patterns—even an iron. There are three drawers—two shallow ones for threads and thimbles, a deeper one for larger accessories. The drawer bottoms protrude at each side to run into dadoes, avoiding hardware or cleats.

But you don't have to sew to make use of this project. It's an ideal hobby center for many activities. All parts come out of two sheets of plywood of different thicknesses. The ⅜″ panel gives you the doors and back plus all the small parts for drawers and partitions—all these parts are coded in the materials list to match

Lester W. Scheuermann
Upper Marlboro, MD

the letter codes on the cutting diagram. There are even parts for a sewing box that features a compartmented tray and a pincushion top. You'll note that the center of the lid is cut away so you can pad it and cover it with material before gluing and bradding it back in place. A "false-top" panel is then set in against this reassembly, as shown with dotted lines.

PANEL LAYOUTS

¾" × **4'** × **8'**

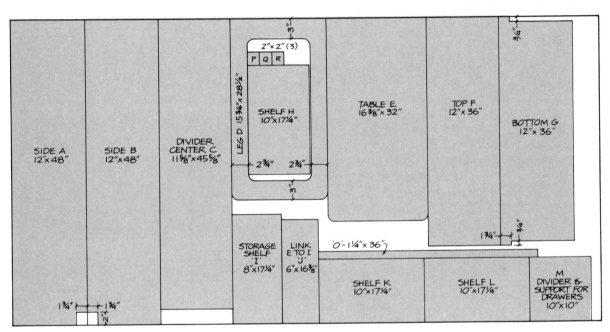

⅜" × **4'** × **8'**

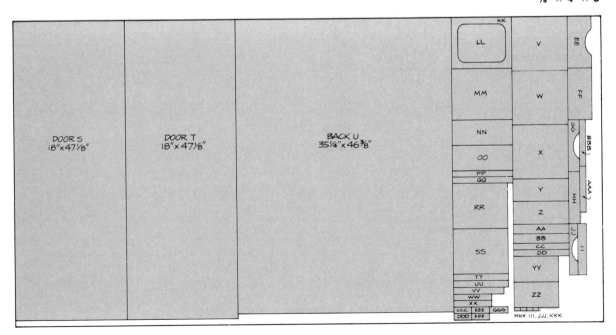

■ MATERIALS LIST ■

Quantity	Description
1 4 x 8 panel	⅜" plywood, MDO, A-C Exterior, or, if available A-B Interior
1 4 x 8 panel	¾" plywood of similar grade
3 pair	Wrap-around hinges (for doors), or other type of hinges that will preclude having to fasten into edge grain of sides A and B
1 pair	Invisible link stop hinges (for fastening J to I)
1 pair	Combination hinge-supports (for fastening E to J)
1 pair	1½" butt hinges (for sewing box lid)
2	Magnetic door catches
As desired	Knobs (for hangers), dowels (for spools, etc.), pin-cushion (for sewing box)
As required	4d finishing nails, white or urea-resin glue, wood dough or synthetic filler, fine sandpaper, top quality latex paint

Note: Dimensions not given with drawings of smaller plywood pieces are as follow (all measurements below in inches):

Drawers:

V, W, X	Bottoms for all 3 drawers—9 x 8⅛
Y, Z	Lower drawer sides—9 x 3½
AA, BB	Middle drawer " —9 x 1½
CC, DD	Top drawer " —9 x 1
EE, FF	Lower drawer front and back—3⅞ x 8⅛
GG, HH	Middle drawer " " " —1⅞ x 8⅛
II, JJ	Top drawer " " " —1⅜ x 8⅛

Sewing Box:

KK	Top—10 x 8
LL	Cutout of KK for pincushion—8 x 6
MM	Bottom—10 x 8
NN, OO	Front and back—10 x 4
PP, QQ	Lid front and back—10 x 1
RR	Inside of lid—9⅛ x 7⅛
YY, ZZ	Sides—7¼ x 4
HHH-KKK	Tray supports—½ x 1

Tray:

SS	Bottom—9⅛ x 7⅛
TT, UU	Front and back—9⅛ x 1
WW, XX	Sides—6⅜ x 1
AAA, BBB	Lid sides—7¼ x 1
CCC-GGG	Dividers—3 x 1

Doors can be equipped with dowels (for hanging scissors and spools currently in use), with a knob for clothes hangers, and with metal plates for the magnetic catches mounted inside the case, as shown in the main drawing.

The hinges for the fold-out table are critical, especially between parts E and J, since these hinges must also function as braces when the table is open. We used a desk-top hinge which has a brace attached to it. The hinge between J and I panels should not protrude on top, so here we used invisible hinges—but knife hinges would work as well. Those angled blocks, formed by slicing part P at an angle (see detail) are mounted inside the corner where D butts E to set the upper leaf of these hinges at an angle. All this is clearest in the side view of the cabinet assembly.

FRONT VIEW

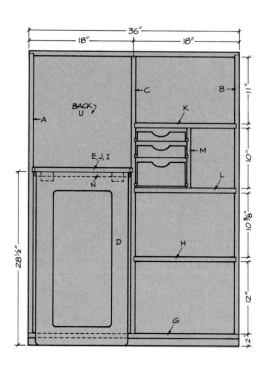

SIDE VIEW & BLOCK DETAIL

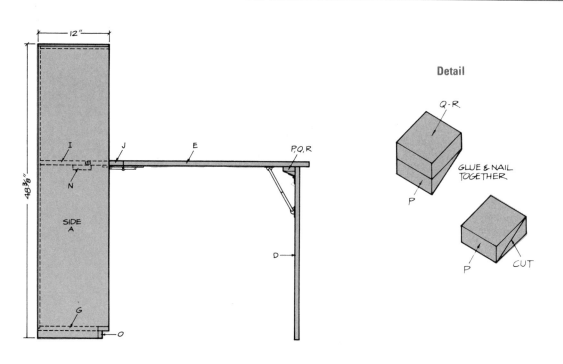

Detail

GLUE & NAIL
TOGETHER

CUT

ISOMETRIC VIEW

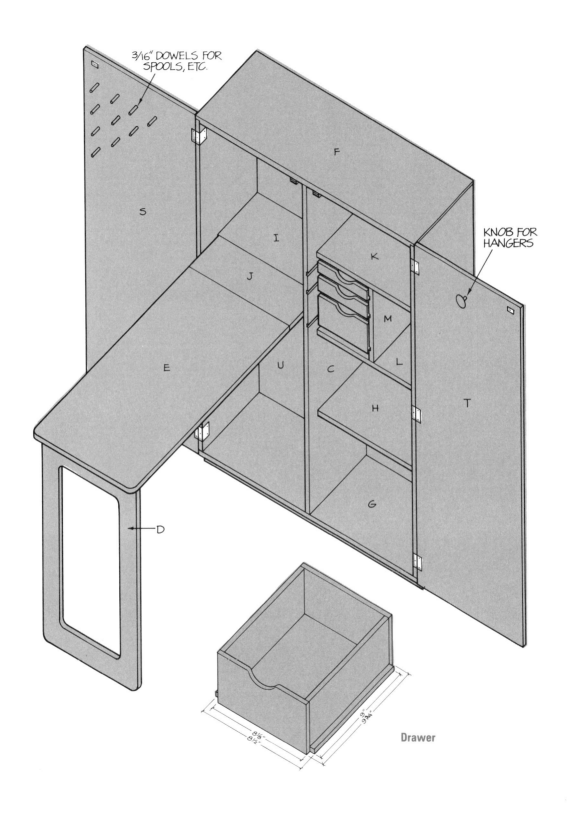

3/16" DOWELS FOR SPOOLS, ETC.

KNOB FOR HANGERS

Drawer

SEWING BOX VIEWS

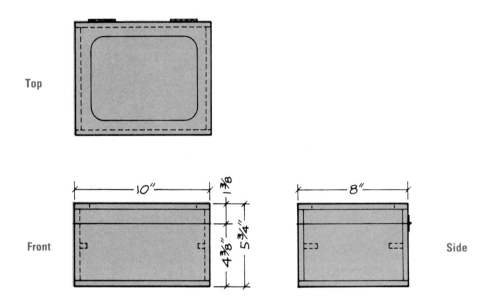

Top

Front

10″

1⅜

4⅜″

5¾″

8″

Side

ISOMETRIC VIEWS

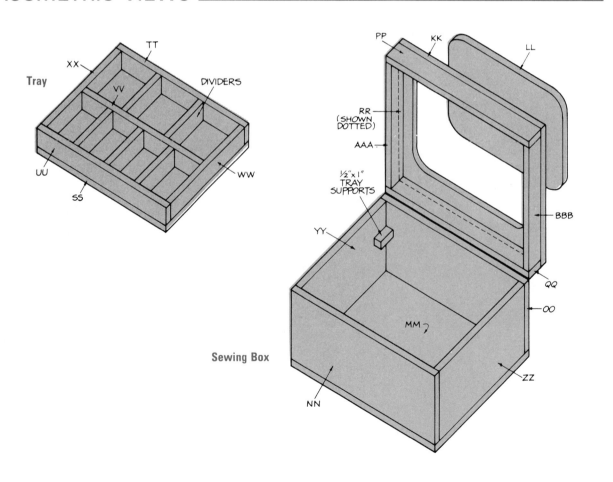

Tray

TT

XX

VV

DIVIDERS

UU

SS

WW

PP

KK

LL

RR
(SHOWN
DOTTED)

AAA

½″ x 1″
TRAY
SUPPORTS

YY

BBB

QQ

OO

MM

Sewing Box

NN

ZZ

VIDEO GAME CENTER

The cutting diagram for this slot-together stand for holding the components of a home computer or video game set is laid out so you could have the full ¾″ sheet cut exactly in half at the lumber yard, for easier transporting—and easier handling once you get it home. If you have the patience to cut all the pieces with a saber saw, you don't need any stationary workshop tools. In fact, this is one of the simplest single-sheet projects in this book, and one that requires almost nothing in addition to the plywood.

Though designed as a "concentration corner" for electronic-game gear (with a bench that will seat two children side-by-side), the unit can double as a home-computer center for an adult (see photo next page). The canted sides of the unit let it tuck into a

Jill Shurtleff
Providence, RI

corner, and whenever space is at a premium, you can pack away the hardware and software and disassemble the stand into units that store virtually flat.

When it's set up, though, it's a sturdy unit, offering a twist-around cord-storage rack and a shelf for game catridges. The only parts that are permanently assembled are the cord-storage rack and the brackets for the three shelves (these are glued and screwed to the inner faces of the side units; because these sides are canted, the shelves are self-positioning and are locked in place by cleverly-located slots and notches).

Since the lines are so simple, you'll want to concentrate on an attractive finish. For our prototype, we chose MDO plywood and filled any edge voids that were revealed by cutting. We sanded the edges carefully and primed them before applying enamel. Our two-tone effect is particularly pleasing: all vertical panels are painted tangerine. All horizontals (shelves and bench) are painted a dark yellow. Be certain to cut all slots enough oversize to accept the panels after they are primed and painted on both sides. Note that some of the slots are deliberately oversize (1″ or 1¹⁄₁₆″) so as to accept cross members at an angle. That makes it

PANEL LAYOUT

¾″ × 4′ × 8′

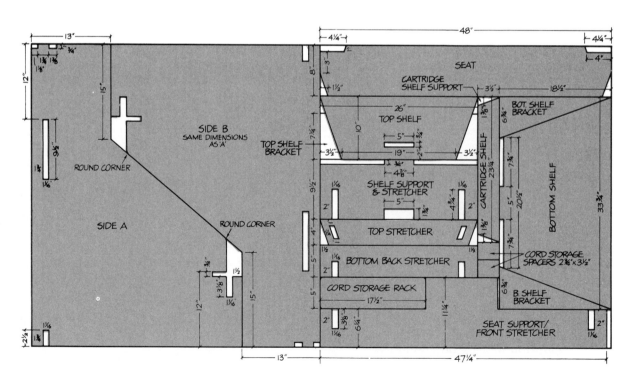

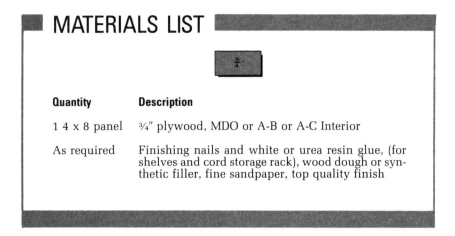

MATERIALS LIST

Quantity	Description
1 4 x 8 panel	¾″ plywood, MDO or A-B or A-C Interior
As required	Finishing nails and white or urea resin glue, (for shelves and cord storage rack), wood dough or synthetic filler, fine sandpaper, top quality finish

unnecessary for you to cut the *slots* at fussy angles—and it doesn't compromise the project's sturdiness when it's set up.

Since both sides of several of the panels will show when the unit is assembled, MDO (with two smooth faces) eases the painting chore. If you prefer a unit with a natural wood finish, you'll have to search for a top-quality A-B Interior Grade panel.

TOP VIEW

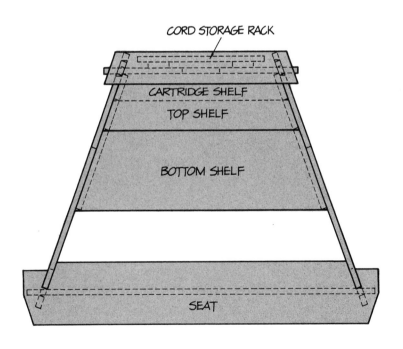

EXPLODED VIEW

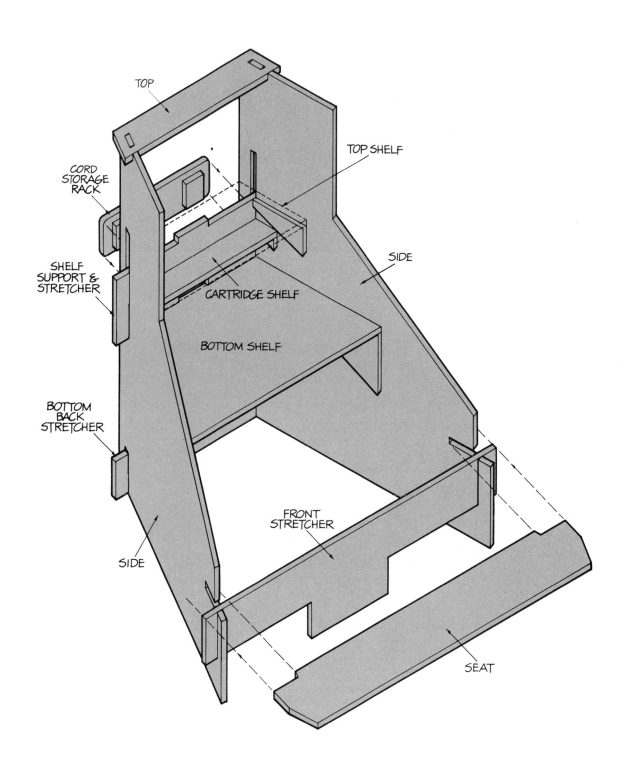

TOP

TOP SHELF

CORD
STORAGE
RACK

SIDE

SHELF
SUPPORT &
STRETCHER

CARTRIDGE SHELF

BOTTOM SHELF

BOTTOM
BACK
STRETCHER

FRONT
STRETCHER

SIDE

SEAT

EXERCISE BENCH

Equipment to keep flab off your body needn't also flatten your wallet. One sheet of ¾″ plywood creates this practical, good-looking bench, and your only other investment will be a set of weights. The bench is sturdy but adjustable, with features of an expensive steel bench—such as leg lift and curl bar, squat rack, and two-position incline back. And the slot-together construction makes the project easy to transport and store.

The pivoting leg bar lets you stack on increased weights, just as you'd do with the barbell; it hangs out of the way when not in use (photo next page). Note the various notches on both sides of the weight rack—including a safety pair to help you get out from under the weight during a bench press.

Greg Sollie
Huntsville, AL

After cutting out all plywood pieces as shown below, cut the 1″ steel pipe into two lengths: 48″ and 26½″. Thread the ends (if you don't have a pipe threader, take the pieces to a plumbing shop). Cut the steel rods to length and bend these to shape with a hammer and vise (turn page for dimensioned detail). Attach slot members to racks and leg-lift braces with glue and right-angle brackets. Laminate the two halves of the bench support with glue and clamps as detailed, and mount the incline supports on the back of the bench with glue and right-angle brackets. Hinge the three units together as shown.

Slot together sides and the front and back cross members, installing the bench incline rods at the same time. Bolt on racks and leg-lift braces. Install the bench seat assembly and the leg lifts and you're ready to sand, prime and paint all parts.

When the paint has dried, upholster the bench seat and back with vinyl material over 1″-thick foam rubber, pulling neat and snug and tacking along the bottom edge. Upholster the leg-lift bars with the same material over 2″-thick foam, but wrap this foam around 1⅛″-dia. cardboard tubing cut to length and capped with 4½″-dia. end disks with a 1¼″-dia. center holes, cut from ¼″ scrap plywood. Machine-stitch a vinyl sleeve to cover, and staple the ends into the plywood disks. Slip these pads onto the two pipes and install the weight-lifting bar collars as shown on the next page. You're now ready to test out the full adjustability of the assembly for your exercise routine.

PANEL LAYOUT

¾″ × 4′ × 8′

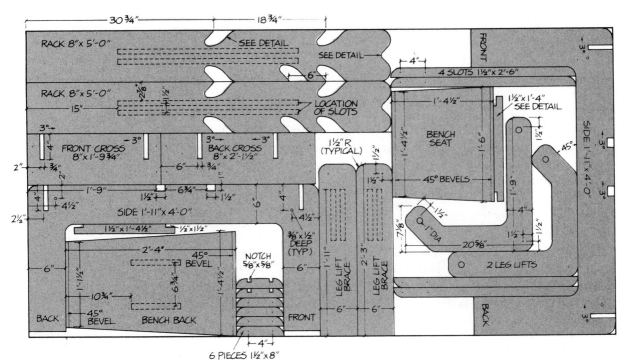

■ MATERIALS LIST

Quantity	Description
1 4 x 8 panel	¾″ plywood, MDO or, if available, A-B Interior
74½ lin. in.	1″ dia. steel pipe
78 lin. in.	⅜″ dia. cold-rolled steel rod
2	1″ dia. threaded steel pipe caps
1 yd.	60″-wide vinyl upholstery material
2 pr.	2½″ x 1″ butt hinges with screws
8	3″ x ¼″ bolts with nuts and washers
2	3″ x ⅝″ machine bolts (with 1½″ shoulders) with 2 lock nuts, 6 washers and 2 finishing caps or acorn nuts
6	1″ inside dia. weight lifting bar collars
16	1″ right-angle brackets with screws
As required	1″-thick foam, 2″-thick foam, 1½″ dia. cardboard tubing, ¼″ scrap plywood or paneling, upholstery tacks, white or urea resin glue, fine sandpaper, top-quality enamel and companion primer

FRONT VIEW: BENCH & LEG LIFT

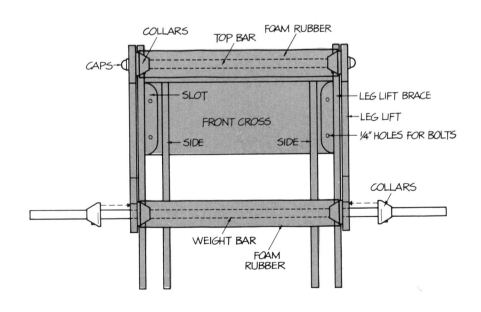

EXPLODED VIEW

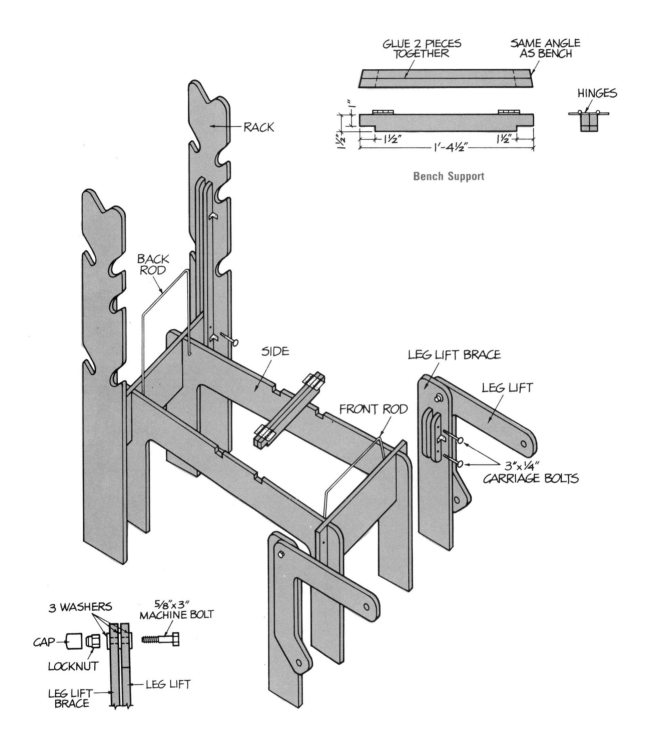

GLUE 2 PIECES TOGETHER

SAME ANGLE AS BENCH

HINGES

1"

½"

1½"

1'-4½"

1½"

Bench Support

RACK

BACK ROD

SIDE

FRONT ROD

LEG LIFT BRACE

LEG LIFT

3"x¼" CARRIAGE BOLTS

3 WASHERS

5/8"x3" MACHINE BOLT

CAP

LOCKNUT

LEG LIFT

LEG LIFT BRACE

BENCH SEAT & SUPPORT

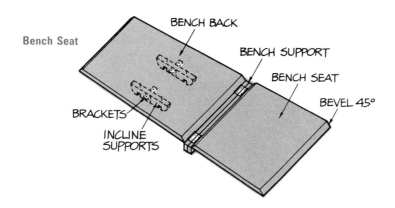

Bench Seat
BENCH BACK
BENCH SUPPORT
BENCH SEAT
BEVEL 45°
BRACKETS
INCLINE SUPPORTS

ROD BENDING DETAILS

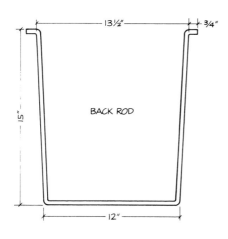

13½"   ¾"

15"

BACK ROD

12"

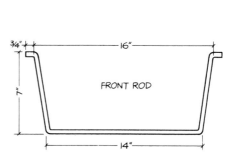

¾"   16"

7"

FRONT ROD

14"

RACK DETAILS

Side

Top

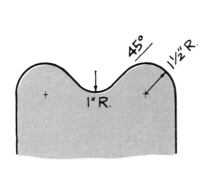

45°   1½" R.

1" R.

Dimensions

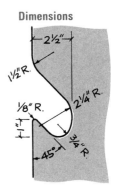

2½"
1½" R.
2¼" R.
⅛" R.
1"
¾" R.
45°

EXERCISE STATION

Like the preceding exercise bench, this unit requires only one sheet of ¾" plywood, but it doesn't even need a set of weights! The muscle-toning tension is built in—by means of (are you ready for this?) vacuum cleaner belts, looped over dowels projecting from the three pivoting arms. One arm press is for use while standing behind the unit, another projects over the bench for use while you're lying down. Since these two arms work independently, you and a workout buddy can do arm presses at the same time. And you can increase the tension by adding more vacuum cleaner belts, in pairs or triplets.

This slot-together, peg-together unit is easily disassembled for storage. Padding of the bench is optional (see previous project

William S. Hoopes
Cohasset, MA

for directions). The parts are so cleverly nested that when you cut them out you'll have only sawdust for waste (virtually). Few dimensions are critical—just be certain all matching parts, such as arms and legs, are identical (note that these can be stacked and drilled a pair at a time). Space has not been left in the layout for saw kerfs, so center your blades on the layout line (except when cutting slots, which should be only slightly oversize to take the ¾" thickness plus the finishing coats you apply to all pieces. In our prototype, we painted all surfaces but the bench top a bright green, and enameled all edges, pegs and dowels a light gray.)

The side and back views show exactly how the vacuum cleaner belts are looped over dowels to tension all three pivoting arms. The series of holes drilled in the uprights let you pass the belts through and lock that loop by inserting a length of dowel. The holes give you great flexibility in positioning the arms at various heights. Note that the pivot rods themselves are short lengths of 1" pipe, threaded on both ends. These "nipples" are available from any plumbing supply and you buy a pair of 3" lengths plus one 10" length for the leg press. The pipes are held in place by turning threaded caps onto each end after they're inserted through the plywood parts.

Dowels an inch in diameter are centered through the doubled arms to serve as operating handles. They're 2' long, and you may want to pad the one in the leg press to keep from "barking" your shins. Note that this pivoting unit is centered at the end of the bench by face-laminated spacers cut from the plywood. Thicker lumber can be substituted, of course.

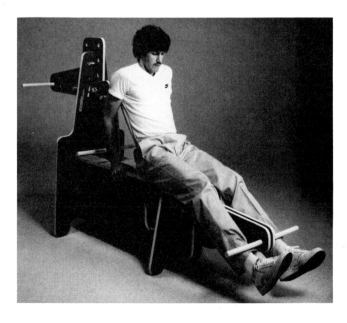
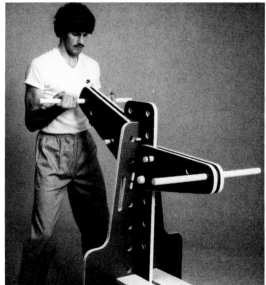

PANEL LAYOUT

¾" × 4' × 8'

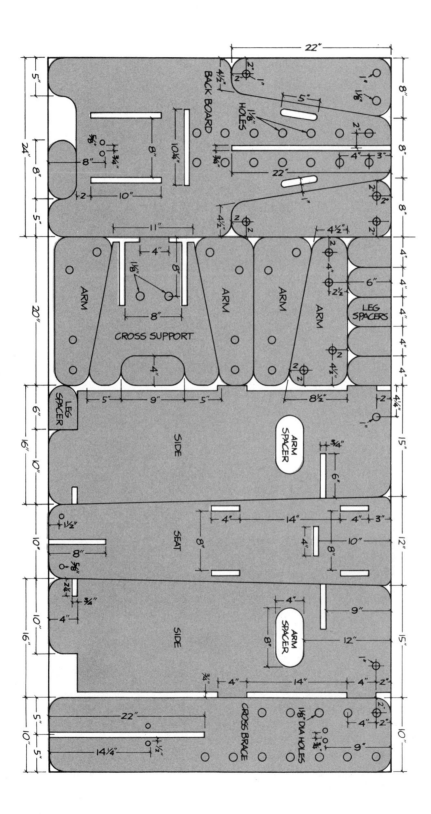

MATERIALS LIST

Quantity	Description
1 4 x 8 panel	¾″ plywood, MDO or A-B or A-C Interior
3	1″ dia. x 24″ wood dowels
3	1″ dia. x 6″ wood dowels
11	⅝″ dia. x 3″ wood dowels
2	1″ dia. x 3″ pipe nipples, each w/2 caps
1	1″ dia. x 10″ pipe nipple w/2 caps
8-12	4″ dia. vacuum cleaner belts (used in pairs or threes according to resistance desired)
As required	Foam cushion and naugahyde for bench cover (or as desired), foam tubing for leg dowel (or as desired), finishing nails (for laminating arm and leg spacers) white or urea resin glue, wood dough or synthetic filler, fine sandpaper, top quality finish

SIDE & BACK VIEWS

Side

Back

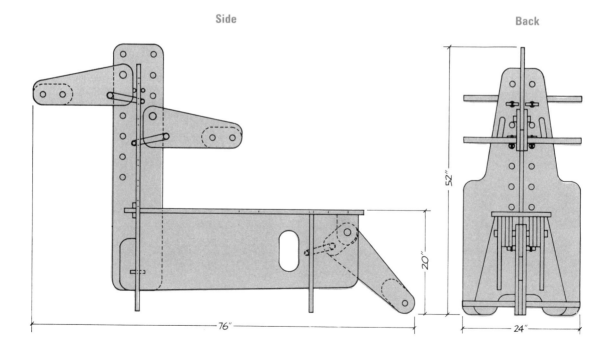

ISOMETRIC VIEW

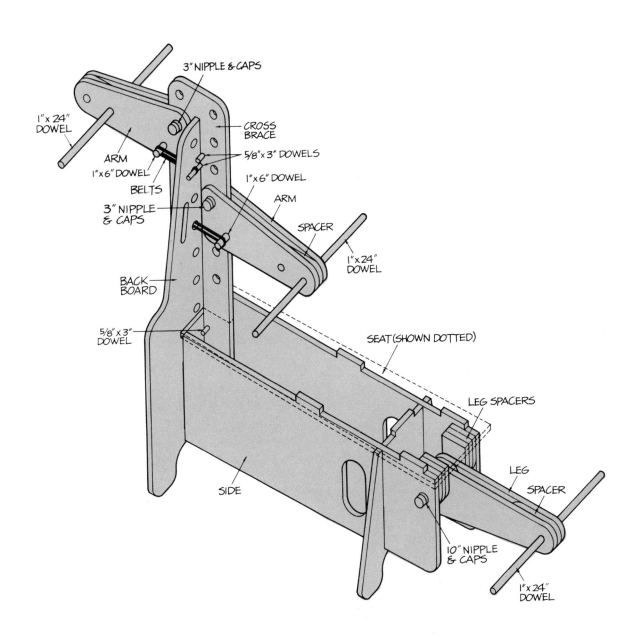

1" x 24" DOWEL

3" NIPPLE & CAPS

CROSS BRACE

ARM

1" x 6" DOWEL

BELTS

5/8" x 3" DOWELS

3" NIPPLE & CAPS

1" x 6" DOWEL

ARM

SPACER

1" x 24" DOWEL

BACK BOARD

5/8" x 3" DOWEL

SEAT (SHOWN DOTTED)

LEG SPACERS

SIDE

LEG

SPACER

10" NIPPLE & CAPS

1" x 24" DOWEL

POTTERS' KICK WHEEL

William Zettel
Tonawanda, NY

For anyone serious about taking up pottery-making, this project's a real kicker. It's highly functional but very compact. And you make the kick wheel yourself from packaged concrete mix.

Lay out all parts on one sheet of ¾″ plywood (MDO preferred, if you're planning to paint the unit). Keep your saber saw blade centered on the layout lines to equalize the kerf on all parts. (The angle and height of the seat can be modified to suit the user, but it's difficult to calculate at this stage; the dimensions shown proved comfortable for a potter of average stature, leaving good turntable clearance.)

Draw an 11″-dia. circle centered on the wheel base (part *H*) and mark locations for the eight equally spaced sheet metal tie screws. Drive screws into the plywood to a depth of ⅝″; the heads protruding from the wood will anchor the concrete to your wheel base. To cast this flywheel: Drill pulley wheels for 6″ stove bolts (see turntable/flywheel detail). Set the wheelbase level on three blocks (bricks or 2×4s on edge) and insert a well-greased shaft (to keep concrete from sticking) through wheelbase to floor. Slide the upper pulley wheel on the shaft and let it rest on the wheelbase. Securely brace the shaft so it is precisely perpendicular to the wheelbase.

Space pulleys on the shaft using set screws (5¼″ centerline to centerline, with bottom pulley flush beneath wheelbase); install stove bolts. Tack flashing around the perimeter of the wheelbase to create a concrete form; fill this to a 2″ depth with concrete.

Center another strip of flashing in an 8″-dia. circle around the shaft, on top of the wheelbase concrete. Push this into the concrete about ⅛″ and pour a 2½″-deep hub. (This hub concrete will nearly cover the upper pulley wheel.) Loosen the pulley wheel set screws and make certain the shaft will turn as the concrete begins to set. As it sets up, remove the flashing and round the top edges of the concrete with a trowel.

Glue-screw the plywood unit together as shown in the drawings. Finish with a good primer and exterior enamel. Mount the kick wheel/turntable assembly. Place the shaft through the bearing in part C and align the shaft plumb by shifting the bearing block

PANEL LAYOUT

¾″ × 4′ × 8′

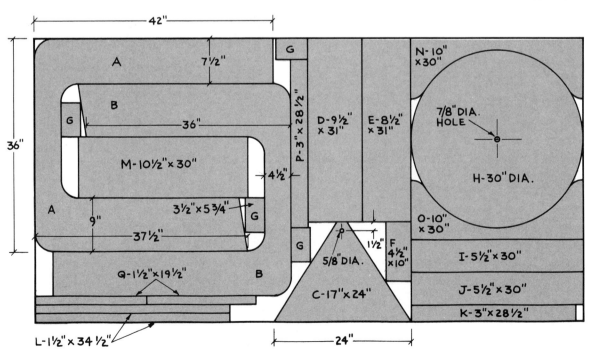

MATERIALS LIST

¾"

Quantity	Description
1 4 x 8 panel	¾" plywood, A-B or A-D Interior, A-B or A-C Exterior, or MDO
1 bag	Concrete mix, 1 cu. ft, for wheel
8 lin. ft	Aluminum flashing, or equivalent, 3" wide, for wheel form
60-70	Wood screws, flathead, #8 x 1½" for glue-screw assembly
8	Screws, sheet metal, #10 x 1½" for wheelbase/concrete tie
2	Stove bolts, ¼" dia. x 6" (with washers) for pulley/kick wheel assembly
2	Lag bolts, ¼" dia. x 3½" (with washers) for bearing block assembly
8	Stove bolts, ¼" dia. x 2" (with washers) for attaching footrests
1	Thrust ball, ⅝" dia. (local hardware)
1	Metal disc, ⅝" dia., for thrust ball mount
2	Ball bearings, flanged, ⅝" (Cat. No. 4X728, W.W. Grainger Co., 5959 W. Howard St., Chicago, IL 60648, or equivalent)
2	Pulley wheels, 6" dia. (Cat. No. 3X919, W.W. Grainger Co., or equivalent)
1	Steel line shaft, ⅝" dia. x 36" long (local hardware or Sears Cat. No. 9H 2830C, cut to length)
1	Sanding disc (throwing head), 8" (Cat. No. AD85-C13, Silvo Hardware Co., 109 Walnut St., Philadelphia, PA 19106, or equivalent)
1 (optional)	Plastic wash basin, convenient size, for splash guard (with ⅝" collar to hold in place)
As required	Waterproof glue to glue-screw assembly; wood dough or other filler for filling countersunk screw holes and plywood edge voids; fine sandpaper for smoothing plywood cut edges and filler; exterior primer and enamel for finishing

assembly. Attach this assembly to base *M* with 1¼" wood screws. Set the kick wheel to the correct height on the shaft and lock in place with set screws. You may want to nick the shaft with a file at appropriate locations for the kick wheel set screws—to increase their holding power.

SIDE & BACK VIEW

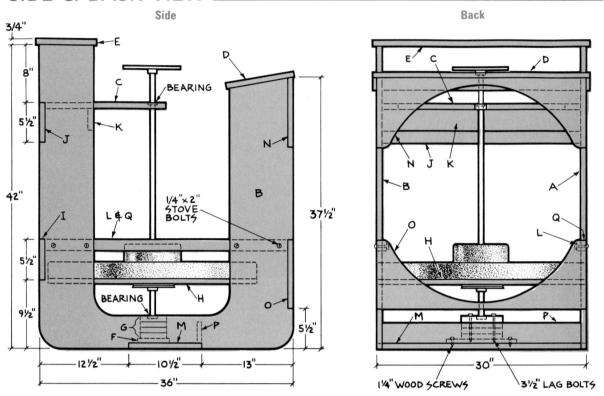

Side

Back

3/4"

E

8"

C

BEARING

5½"

K

J

42"

I

L & Q

1/4"×2" STOVE BOLTS

BEARING

H

G

F

M

P

12½"   10½"   13"

36"

D

N

B

37½"

O

5½"

E   C   D

N J K

B   A

O   H   Q

L

M   P

30"

1¼" WOOD SCREWS   3½" LAG BOLTS

9½"

5½"

TURNTABLE ASSEMBLY DETAIL

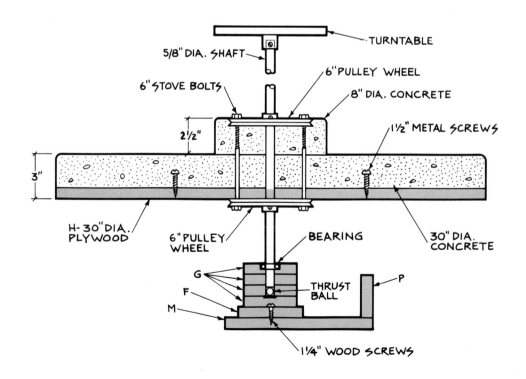

5/8" DIA. SHAFT

TURNTABLE

6" STOVE BOLTS

6" PULLEY WHEEL

8" DIA. CONCRETE

1½" METAL SCREWS

2½"

3"

H- 30" DIA. PLYWOOD

6" PULLEY WHEEL

BEARING

30" DIA. CONCRETE

G

F

M

THRUST BALL

P

1¼" WOOD SCREWS

ISOMETRIC VIEW

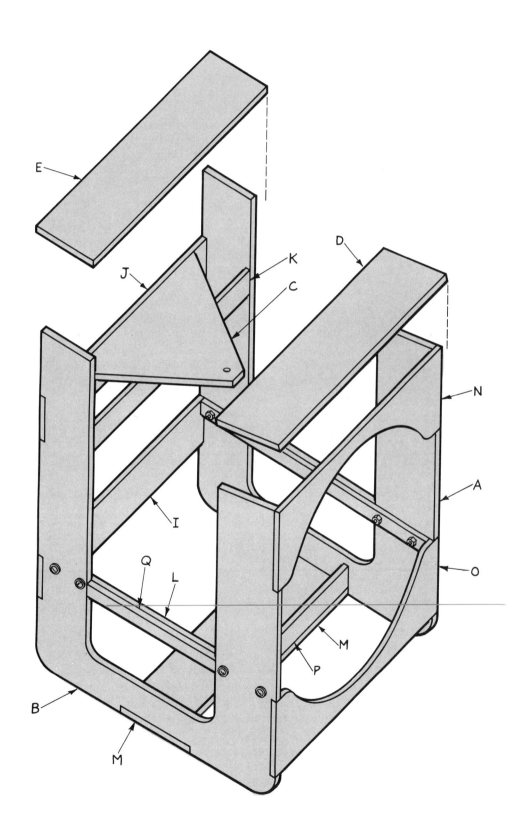

DOUBLE BUNK/DIVIDER

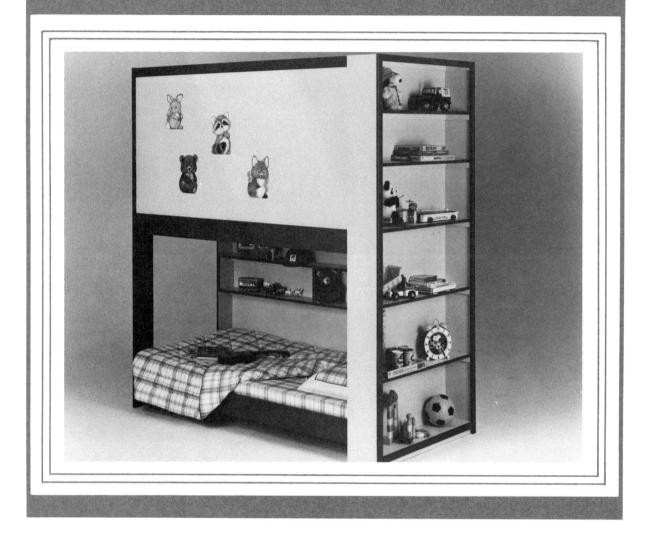

A long, skinny room can be effectively shared by two children as a bedroom when you center this unit against one long wall. The upper and lower bunks face in opposite directions for privacy—and to discourage after-hours chatter. Even the bookshelves at the end can be apportioned—three apiece; each bunk has its own private storage shelves built in, too, and all amenities are scrupulously equal, to avoid territorial disputes.

Built from just six ¾″ panels (no quality less than A-B Interior will do, since both sides of most panels will show), the waste in this shrewdly-designed project is kept to a cost-cutting minimum. If you plan to finish the project with paint, you're better off choosing MDO plywood. Our prototype was two-toned—bright

Greg Branch
Salinas, CA

yellow trimmed with a warm brown; a few sprightly decals dress up the large "billboard" surfaces, but you could substitute supergraphics, perhaps on the theme of the bunk-mates' names.

Dowels of large diameter (we used 1½") form the ladder rungs and the guard rail for the upper bunk. Note that the rail "breaks" at the ladder for easier climbing access. This feature also makes it easier to change the linen on this mattress—and to tuck in that elevated sleeper. All dowel ends are glued and bradded into holes drilled through the ladder sides—and at the shaped top of vertical support Z. But the right end of the short rail must be supported by a closet-rod bracket or a homemade block.

Since the project bolts together, it can be broken down into

PANEL LAYOUTS

1

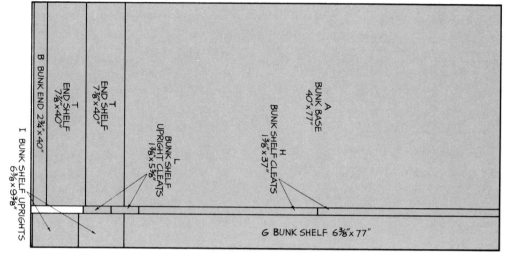

3/4" × 4' × 8'

B BUNK END 2¾"× 40"

T END SHELF 7⅞"× 40"

T END SHELF 7⅞"× 40"

I BUNK SHELF UPRIGHTS 6⅜"× 9⅜"

L BUNK SHELF UPRIGHT CLEATS 1⅜"× 5⅝"

A BUNK BASE 40"× 77"

H BUNK SHELF CLEATS 1⅜"× 37"

G BUNK SHELF 6⅜"× 77"

2

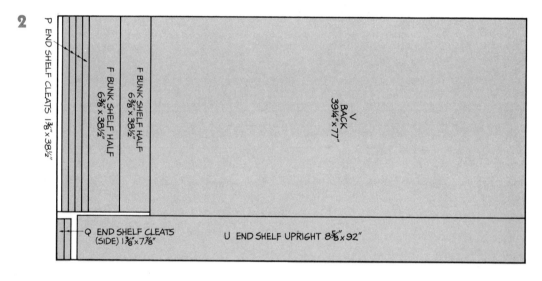

3/4" × 4' × 8'

P END SHELF CLEATS 1⅜"× 38½"

F BUNK SHELF HALF 6⅜"× 38½"

F BUNK SHELF HALF 6⅜"× 38½"

V BACK 39¼"× 77"

Q END SHELF CLEATS (SIDE) 1⅜"× 7⅞"

U END SHELF UPRIGHT 8⅝"× 92"

3

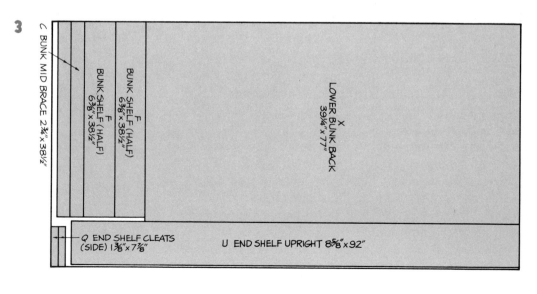

3/4" × 4' × 8'

C BUNK MID BRACE 2¾"× 38½"

F BUNK SHELF ('HALF') 6⅜"× 38½"

F BUNK SHELF ('HALF') 6⅜"× 38½"

X LOWER BUNK BACK 39¼"× 77"

Q END SHELF CLEATS (SIDE) 1⅜"× 7⅞"

U END SHELF UPRIGHT 8⅝"× 92"

PANEL LAYOUTS

4

¾" × 4′ × 8′

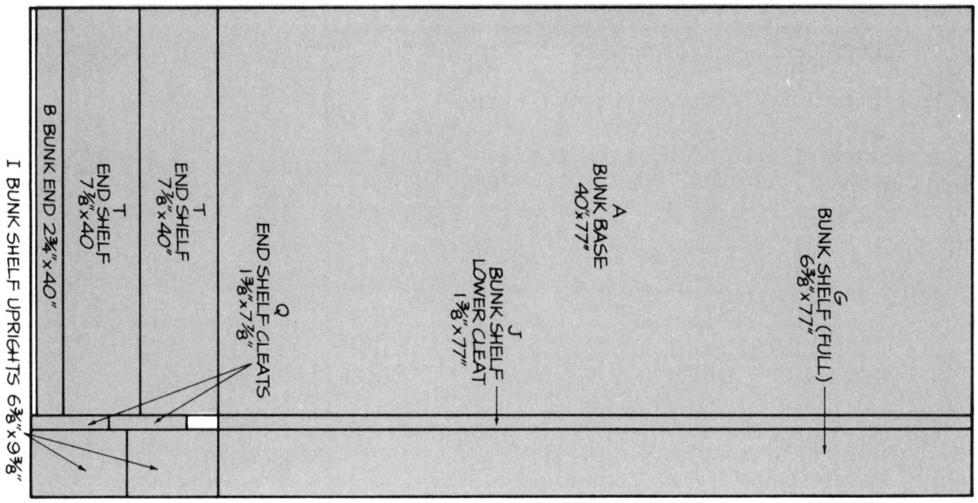
B BUNK END 2¾"x40"
T END SHELF 7⅞"x40"
T END SHELF 7⅞"x40"
Q END SHELF CLEATS 1⅜"x7⅞"
A BUNK BASE 40"x77"
J BUNK SHELF LOWER CLEAT 1⅜"x77"
G BUNK SHELF (FULL) 6⅜"x77"
I BUNK SHELF UPRIGHTS 6⅜"x9⅜"

5

¾" × 4′ × 8′

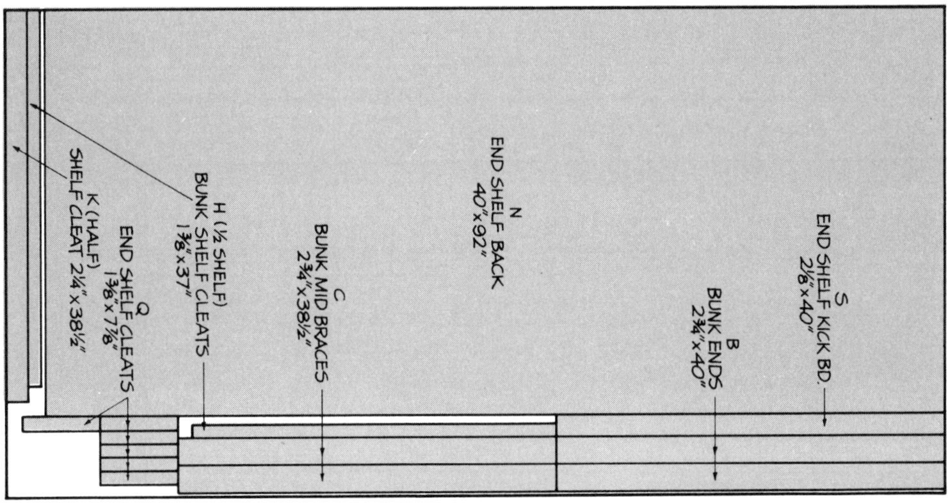
K (HALF) SHELF CLEAT 2¼"x38½"
Q END SHELF CLEATS 1⅜"x7⅞"
H (½ SHELF) BUNK SHELF CLEATS 1⅜"x37"
C BUNK MID BRACES 2¾"x38½"
N END SHELF BACK 40"x92"
B BUNK ENDS 2¾"x40"
S END SHELF KICK BD. 2⅛"x40"

6

¾" × 4′ × 8′

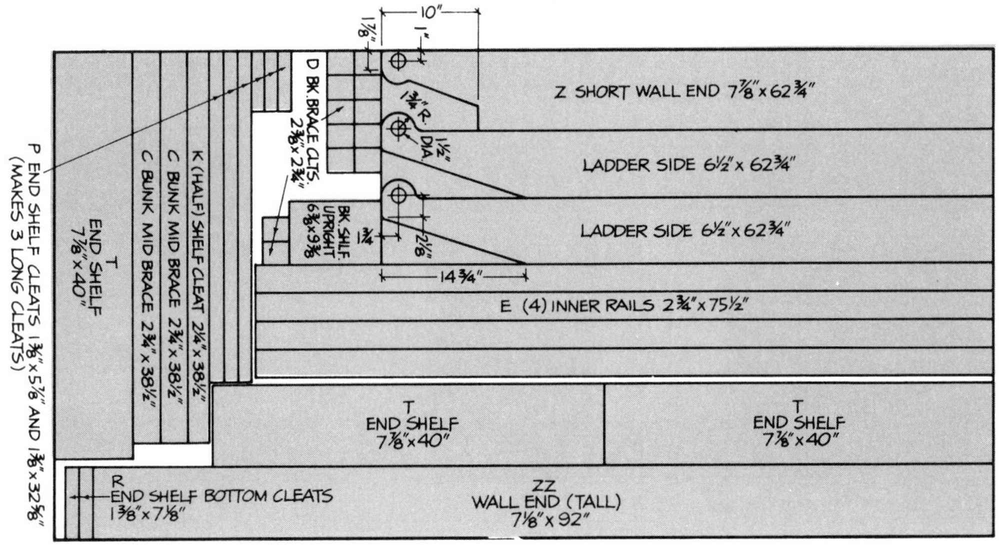
10"
1⅞"
1"
Z SHORT WALL END 7⅞"x62¾"
1¾" R.
1½" DIA.
LADDER SIDE 6½"x62¾"
LADDER SIDE 6½"x62¾"
D BK. BRACE CLTS. 2⅞"x2¾"
BK. SHLF. UPRIGHT 6⅜"x9⅜"
1¼
2⅛"
14¾"
E (4) INNER RAILS 2¾"x75½"
P END SHELF CLEATS 1⅜"x5⅞" AND 1⅜"x32⅝" (MAKES 3 LONG CLEATS)
T END SHELF 7⅞"x40"
C BUNK MID BRACE 2¾"x38½"
C BUNK MID BRACE 2¾"x38½"
K (HALF) SHELF CLEAT 2¼"x38½"
T END SHELF 7⅞"x40"
T END SHELF 7⅞"x40"
R END SHELF BOTTOM CLEATS 1⅜"x7⅛"
ZZ WALL END (TALL) 7⅛"x92"

MATERIALS LIST

Quantity	Description
6 4 x 8 panels	¾" plywood, MDO or A-A or A-B Interior
24	5⁄16" dia. x 2" long bolts with washers and nuts
As required	Finishing nails, white or urea resin glue, wood dough or synthetic filler, fine sandpaper, top quality finish

Key No.	Quantity	
1	4	1" x 6" x 77" lumber rails
2	2	1" x 2" x 92" border trim
3	1	1" x 2" x 38½" border trim
4	1	1" x 2" x 41½" border trim
5	6	1" x 2" x 10⅞" book shelf trim
6	4	Galvanized sheet metal angle clips with screws
7	1	Closet rod bracket for 1½" dowel
8	11 lin. ft.	1½" dia. dowel for ladder runs and guardrail

FRONT VIEW

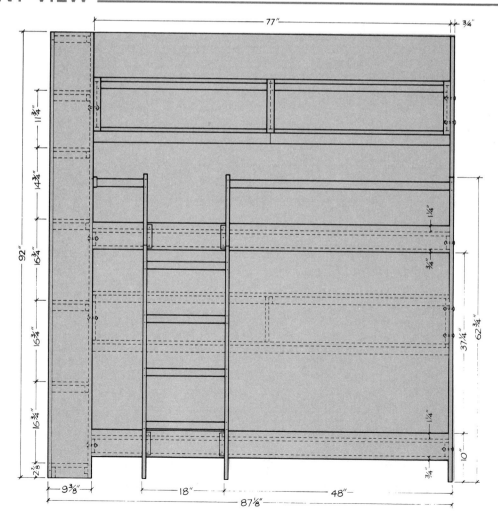

modular components for easy portability (see exploded view). This feature also gives you the option of facing both bunks in the same direction, in case the unit is not needed as a divider and must be backed up against a wall.

Unlike many of the projects in this book, the bunk-bed's cutting diagrams were devised to allow for an eighth-inch saw kerf on every line. So the panel dimensions shown are the actual size you'll end up with if you make your cuts with a blade of that thickness. Make all permanent butt joints with glue and finishing nails. You then enlist a helper for a "trial assembly" of the components to determine the most effective locations for inserting assembly bolts. Drill holes in these locations, through one of the components, then reposition it so exactly-mating holes can be drilled through the second component, using the holes in the first as a template. You'll be delighted to find how sturdy the whole unit is once all bolts are in place.

SIDE VIEW

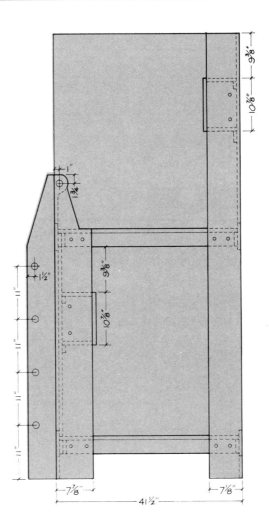

EXPLODED VIEW: FRONT

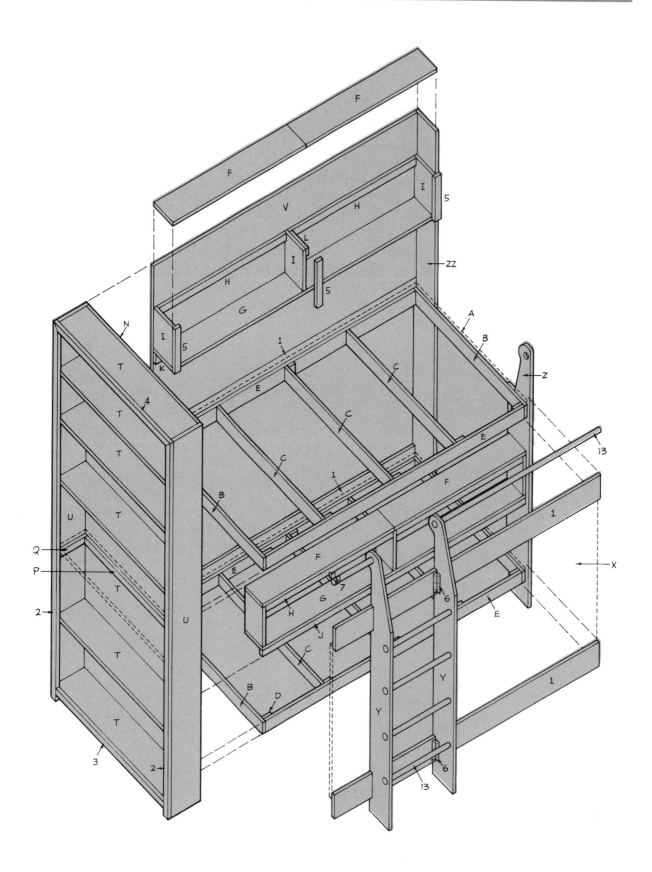

ISOMETRIC VIEW: BACK

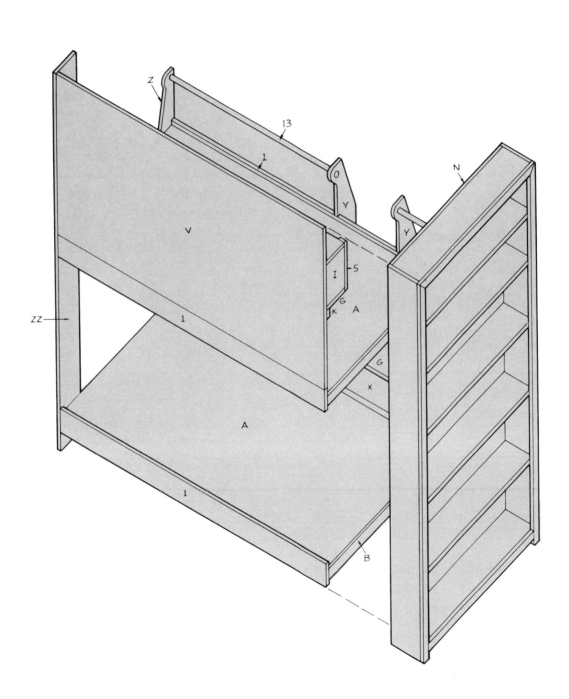

BUNK/DESK/PLAYHOUSE

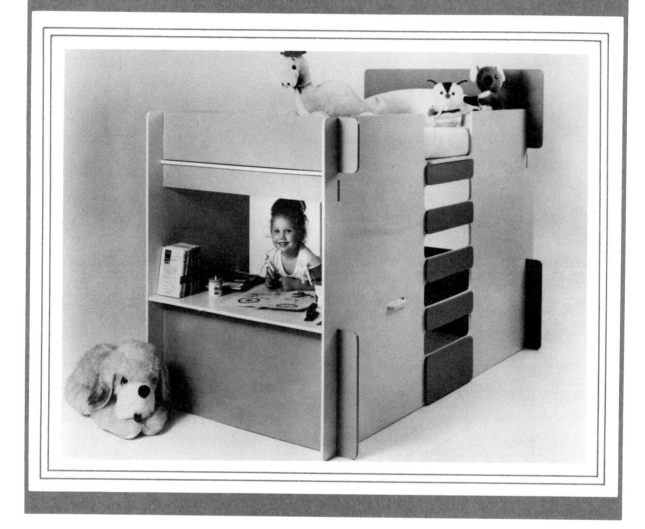

Here's as ingenious a design for a child's bed and desk as you'll find anywhere: The parts slip together, so you don't need nails or screws for assembly—and the unit doubles as a delightful playhouse. Measuring about 6½' in length and 5' high, it's under 4' wide and can be built from just three sheets of ½" plywood. Since the door cutout becomes the desk top, and the step-hole cutouts are used to double-up the thickness of the ladder steps (to make the climb easier on bare feet), there's very little waste. And best of all, the multi-purpose unit shortens the trip from playtime to bedtime.

The unit is scaled to take a standard youth-size foam mattress—and to put it at just the right height for no-stoop bed making. And if you build it all from MDO plywood, for easy painting, you can duplicate the smashing two-color effect we got on our

Donald A. Maxwell, Jr.
Richmond, VA

PANEL LAYOUTS

½" × **4'** × **8'**

½" × **4'** × **8'**

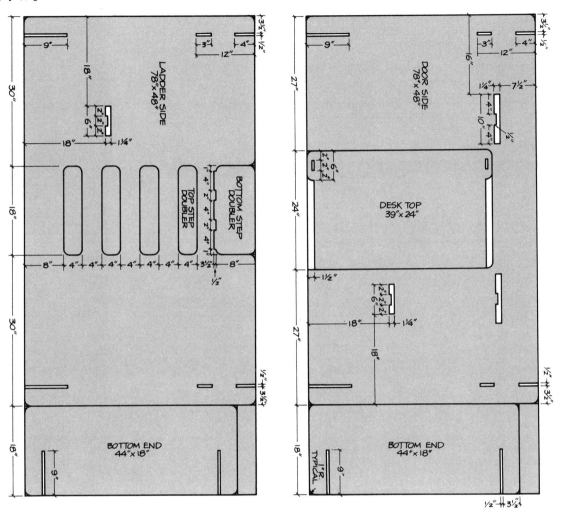

½" × **4'** × **8'**

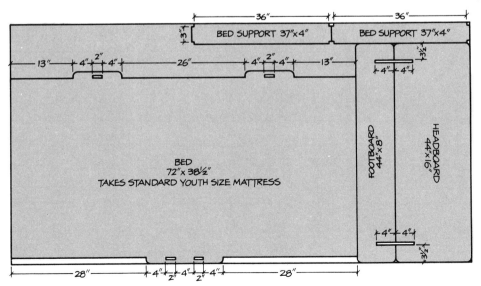

MATERIALS LIST

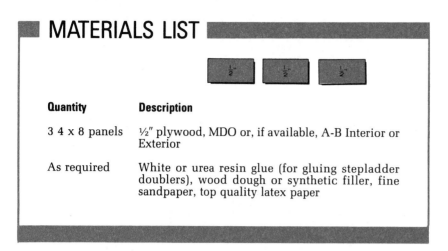

Quantity	Description
3 4 x 8 panels	½″ plywood, MDO or, if available, A-B Interior or Exterior
As required	White or urea resin glue (for gluing stepladder doublers), wood dough or synthetic filler, fine sandpaper, top quality latex paper

ISOMETRIC VIEW: DOOR SIDE

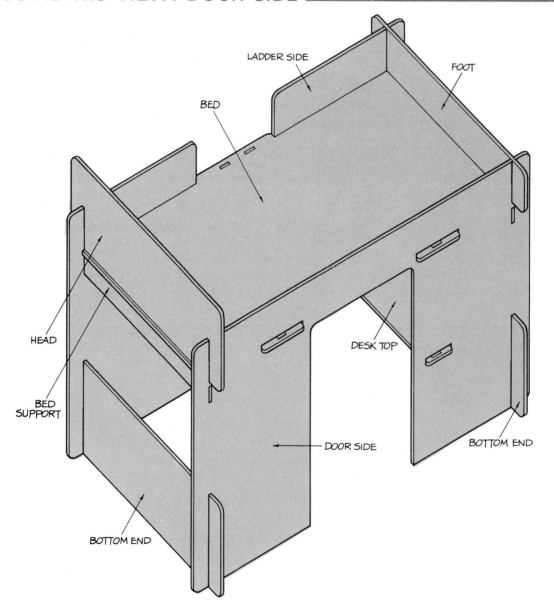

prototype, using daffodil yellow on the "walls" and a cool medium blue on the slot-in cross members and the ladder steps.

The assembly system has a clever tab-lock device: The slots into which the tabs are inserted are 1¼" wide and have centered knuckles that snug into sockets in the tabs. You insert the tab at the top of the slot, then drop it over the knuckle to lock it in place. You'll have to do a lot of clearance-hole drilling to position your sabre-saw blade for cutting all those slots. And when you come to the step cut-outs, either drill an overlapping series of holes no wider than the kerf the blade will leave, or try for a neat plunge cut. You want to keep the cutout piece whole and un-notched. When all cutting's done, sand all edges and check for any voids that should be filled with wood dough (or a small glue block, if quite long) before you prime for painting. Finish all parts before assembly—even the ladder doublers, which can then be fastened in place with glue and finishing nails.

ISOMETRIC VIEW & DETAILS: LADDER SIDE

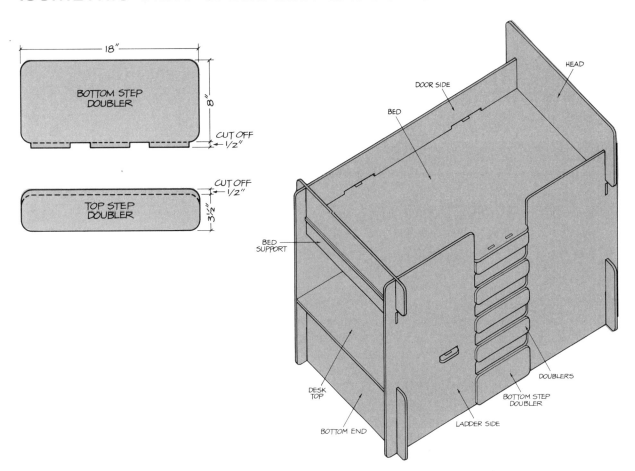

'PLAY CENTER/GUEST BED

Your child may enjoy a surge in popularity if you add this unit to the playroom or den. Young guests will flock to its play facilities. There's a detachable desk that adjusts to two heights by simply tipping it out of its shaped side brackets. You can lift the desk free to use as a lap tray—or as a floor easel, for those artists who prefer to paint while lying on their stomachs. Above the desk is a cork board for posting the masterpieces thus created. At the opposite end there's an ample chalkboard for more transitory creations. Between these art zones there's a literature rack— a deep pocket for favorite story and coloring books—plus cubbyholes for small creatures, blocks and other play stuff.

This entire unit is hinged to the wall, so if a playmate becomes

Dave Fullmer
Salt Lake City, UT

so engrossed in a project he or she can't bear to depart, the unit swings down as a bed platform. Only a few of the stored items need to be removed from its underside before you unhook the top corners and lower the unit, latching its hinged foot-assemblies. The removed items can be stored in that box-like desk seat. The hinged lid doubles as seat and night stand. All parts for both assemblies can be cut from two sheets of plywood .

Note that the tapering end members are dadoed down their inside centers to take the ½"-thick deck panel. The same is true of both the narrow (top) and wide (bottom) bed rails, which project into rabbets cut across the top and bottom of the sides. The cubbyholes are formed with a half-slot "egg-crate" assembly which is then glued and nailed between the vertical dividers.

The holes drilled through the desk top are optional and can be sized to take drop-in jars of poster paints. Note that bumper tacks are driven into the under edge of the desk supports so they won't scar the chalkboard when the desk is removed from its

MATERIALS LIST

Quantity	Description
1 4 x 8 panel	¾" plywood, MDO or, if available, A-B Interior
1 4 x 8 panel	½" plywood, similar grade
72 lin. ft	1½" piano hinge
2	Snap hooks
2	Large screw eyes
2	Small screw eyes
2	Pull catches
1 pr.	2" x ¾" pin hinges
2	2½" x ¼" carriage bolts with nuts and washers
1	16" x 25" corkboard
6	Small bumpers
Approx. 12 lin. in.	Safety chain
As required	Assorted nails and wood screws, white or urea resin glue, wood dough or synthetic filler, fine sandpaper, blackboard panel, top-quality latex paint and companion primer.

side brackets and stored in that compartment (tucked behind desk-retainer blocks) while the bed is lowered.

We painted our prototype in two colors: A deep blue with lighter green for striping around the bed rails and across each box front. The "blackboard" is actually green as well: The MDO plywood surface here is coated with a chalkboard paint.

PANEL LAYOUTS

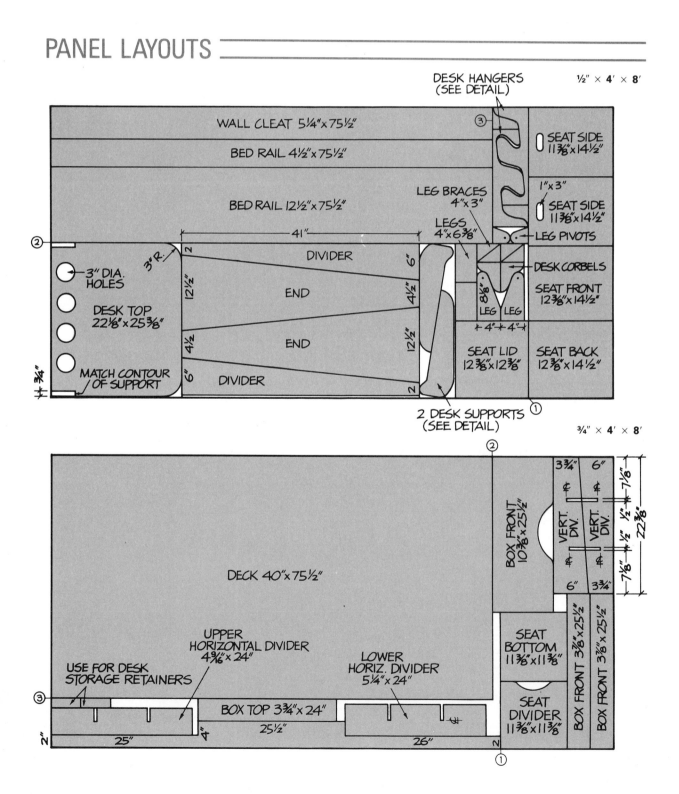

EXPLODED VIEW

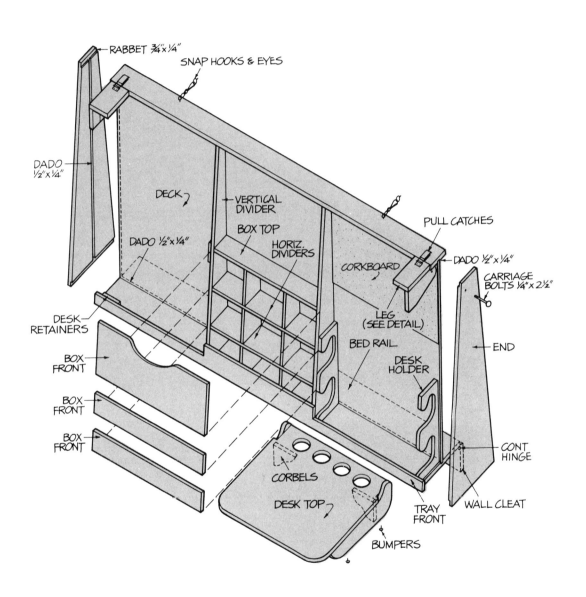

RABBET 3/4" x 1/4"

SNAP HOOKS & EYES

DADO 1/2" x 1/4"

DECK

VERTICAL DIVIDER

BOX TOP

DADO 1/2" x 1/4"

HORIZ. DIVIDERS

PULL CATCHES

CORKBOARD

DADO 1/2" x 1/4"

CARRIAGE BOLTS 1/4" x 2 1/2"

DESK RETAINERS

LEG (SEE DETAIL)

BED RAIL

END

BOX FRONT

DESK HOLDER

BOX FRONT

BOX FRONT

CONT HINGE

CORBELS

WALL CLEAT

DESK TOP

TRAY FRONT

BUMPERS

CUTTING DIAGRAMS

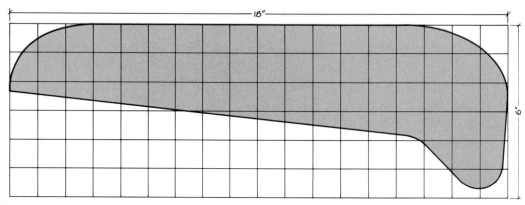

18″

6″

Desk Hangers

Desk Supports

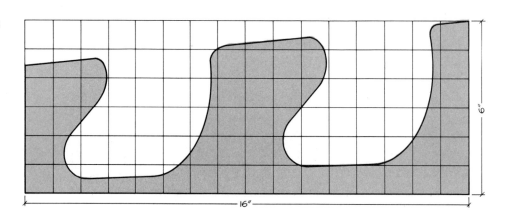

6″

16″

LEG DETAIL   SEAT: EXPLODED VIEW

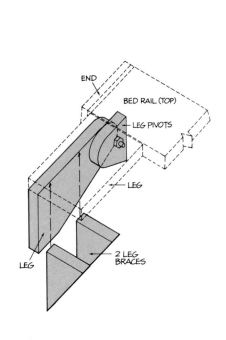

END

BED RAIL (TOP)

LEG PIVOTS

LEG

LEG

2 LEG
BRACES

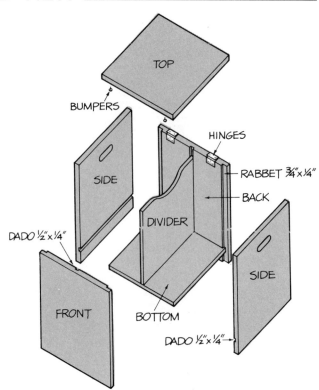

TOP

BUMPERS

HINGES

SIDE

RABBET ¾″×¼″

BACK

DIVIDER

DADO ½″×¼″

SIDE

FRONT

BOTTOM

DADO ½″×¼″

PLAY TOWER WITH SLIDE

It's a climber, a playhouse, a slide—and will keep small children and toddlers occupied for hours at a time. The tower's three walls have long, angled slots to permit assembly into a sturdy triangle that supports the slide. It comes apart easily for storage, or to move the unit outside in good weather.

The tower panels are all 4' across and two of them stand a full 4' tall. The slide is 5½' long, and all parts come out of two panels of ½" plywood. When the unit's not in use, the tower doubles as a storage bin for bulky toys.

Do a complete parts layout on the back face of the panels and cut all pieces with a portable circular saw and saber saw. Glue and screw the platform supports to the sides and do a trial as-

Christopher P. Cork
Oakland, CA

sembly of the three sides, as shown in the corner detail, to see if you must make any adjustments to the angled slots. If you plan to two-tone your project, as we did (a light yellow on large surfaces with a medium blue to accent edges, slide rails and steps), you'll want to paint the parts now, before further assembly. Hinge the door back into its cutout in side *A* and determine whether you wish to hinge the slide to side *C* as shown or devise a quicker knockdown attachment with metal angles screwed to the slide to slip into brackets on the tower.

PANEL LAYOUTS

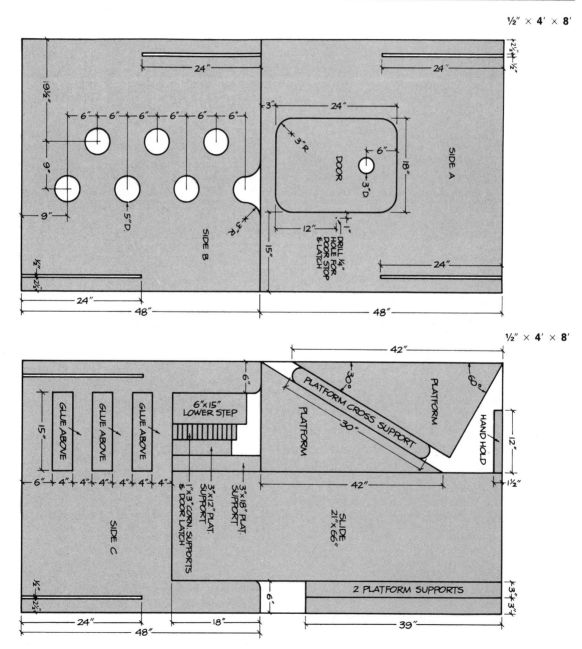

MATERIALS LIST

Quantity	Description
2 4 x 8 panels	½″ plywood, MDO or A-C Exterior, or, if available, A-B Exterior
1 pair	Loose-pin door hinge
6	¾″ #10 flathead machine bolts with nuts (for fastening door hinge to slide)
11 lin. ft	1 x 3 framing lumber (cut to 66″ lengths for side rails)
3	Angle brackets (for platform)
6	2½″ x ¼″ bolts with 12 washers and 6 nuts (for fastening corner supports)
1	2″ x ¼″ bolt with nut and two washers (for fastening door latch)
30	1″ #8 flathead wood screws (for fastening platform supports)
12	1½″ #8 flathead wood screws (for fastening side rails)
1 pair	Butt hinge (for attaching door)
As required	White or urea resin glue, wood dough or synthetic filler, fine sandpaper, top-quality latex paint

TOP VIEW

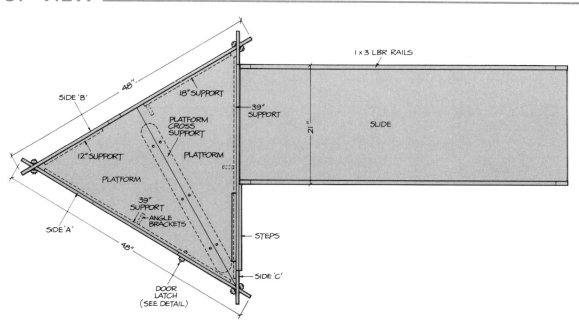

SIDE VIEW

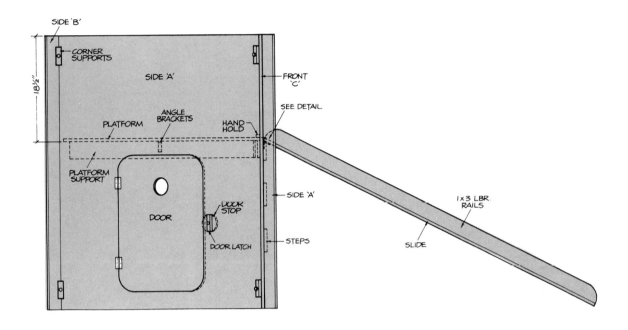

DOOR LATCH DETAIL

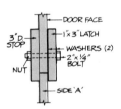

SLIDE ATTACHMENT DETAIL

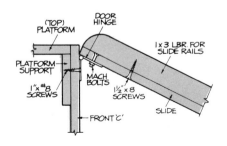

ISOMETRIC VIEW

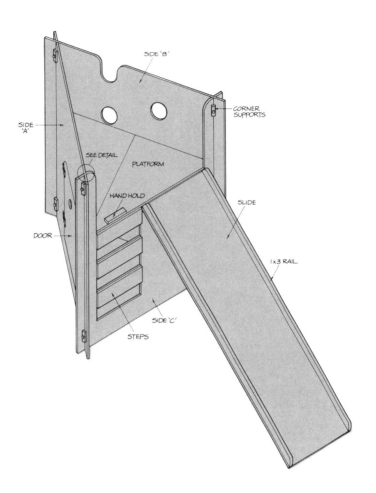

CORNER DETAILS

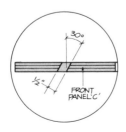

Cutting Dimensions

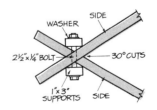

Connection

TOY BOX/BENCH

Would you believe that a sturdy, dual-use project of this size could possibly come from a half sheet of ½″ plywood? The secret lies in the incredibly efficient use of the panel (turn page for layout sketch) and in the ingenious and innovative use of ordinary plastic laminate—the stuff that's made for gluing on counter-tops—as a secondary material: It's soaked in a bathtub of hot water so it can be bent to form the toy trough, and then is clamped between mating members of the end frames. The hinged lid hides the stored toys and doubles as a seat. The end members then become curved arm-rests.

The unit slots together for easy assembly—and that's a big bonus if you wish to ship it to a distant young relative or deliver

Peter Favot
Kanata, Ont., Canada

it personally as a birthday or Christmas surprise. There's almost nothing to it when all parts are stacked flat.

Carefully cut out all parts as dimensioned on the layout drawing, taking special pains with those lower 6¾"-radius cuts (since you're cutting the edges of two parts of the laminate clamp at once). Note that the assembly slots offer the ideal place for drilling a hole to insert your sabre-saw blade.

After painting all plywood parts (our prototype is two-tone: a burnt orange contrasted with bright yellow laminate and a lid painted to match), fasten the end bars across the top of the inner end panels, flush with their top edges, using roundhead screws.

Note that one of the back bars is ½" wider than the other three bars; that's so its top edge will stick up higher, for fastening one leaf of the continuous hinge to (you should avoid driving screws into plywood edges whenever possible). When the lid swings forward onto the double front bars, this ½" discrepancy in height is absorbed by rubber bumpers along their top edge. The sketch shows 1" squares, but you can buy bumper tacks that are easier to apply. The lid needs no restraint chain since in its up position it leans against the back rest, but to avoid an accidental drop on small hands or heads, you may want to add a sliding brace.

Now comes the fun part. Drill mating lead holes through all the bars and drill larger matching clearance holes through the edges of the plastic laminate which you have cut to size. Now soak the laminate in hot water until it becomes pliable. Using the inner end panels as templates, bend the laminate to the desired curvature and sandwich its edge between the pairs of front and back bars. Secure with wood screws.

■ MATERIALS LIST ■

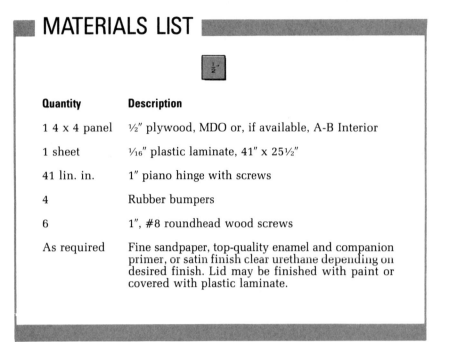

Quantity	Description
1 4 x 4 panel	½″ plywood, MDO or, if available, A-B Interior
1 sheet	¹⁄₁₆″ plastic laminate, 41″ x 25½″
41 lin. in.	1″ piano hinge with screws
4	Rubber bumpers
6	1″, #8 roundhead wood screws
As required	Fine sandpaper, top-quality enamel and companion primer, or satin finish clear urethane depending on desired finish. Lid may be finished with paint or covered with plastic laminate.

PANEL LAYOUT

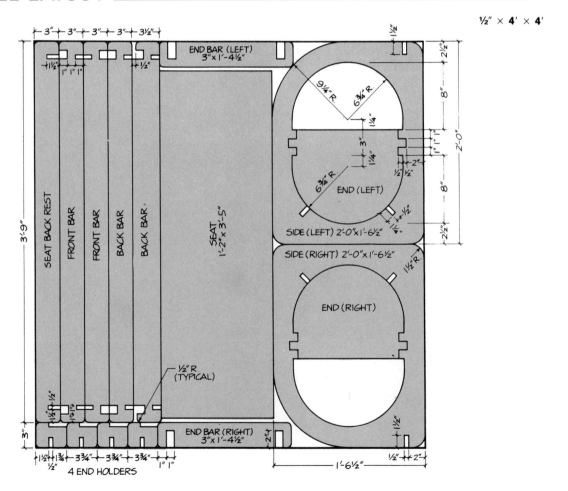

Assemble the slot-together bars and end sections. The laminate thickness should replace the wood removed by the saw kerf in separating the end panels, thus clamping the laminate and preserving its curvature as it cools. The four end holders also support the laminate and prevent any racking of the ends.

The laminate will be strong enough to support the weight of any stored toys, but children should be discouraged from climbing into the trough themselves. But then, it's always a good idea to teach children not to crawl into spaces bigger than they are when there's any door or lid attached.

FRONT VIEW

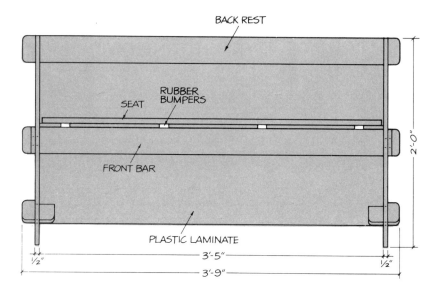

SIDE VIEW

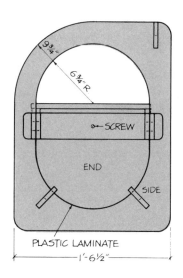

ISOMETRIC VIEW

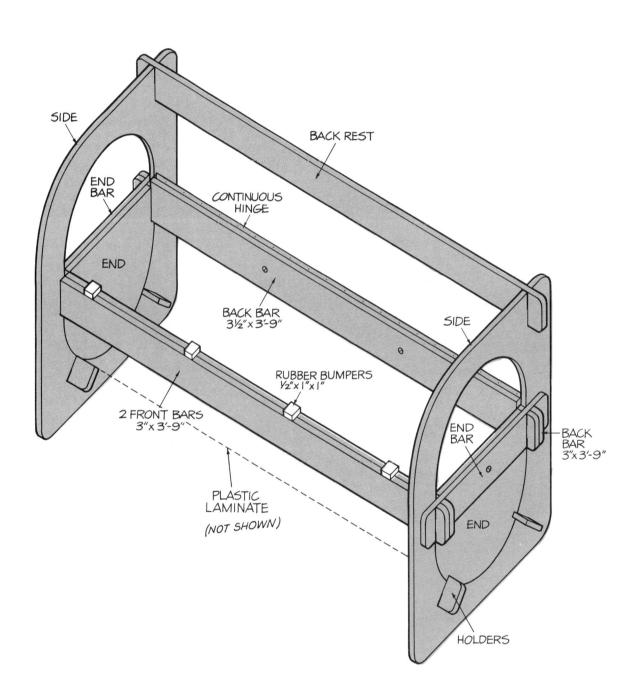

SIDE

END BAR

END

2 FRONT BARS 3" x 3'-9"

PLASTIC LAMINATE *(NOT SHOWN)*

BACK REST

CONTINUOUS HINGE

BACK BAR 3½" x 3'-9"

RUBBER BUMPERS ½" x 1" x 1"

SIDE

END BAR

BACK BAR 3" x 3'-9"

END

HOLDERS

CABOOSE TOY BOX

All aboard! The prizewinning designer of this caboose personalized his project with the name and phone number of the grandchildren he built it for; they often empty out the toys so they can climb in and play train. For safety, you'll want to add sturdy lid supports so the hinged roof sections can't fall accidentally.

The design is clean enough—and detailed enough—to be quite decorative if left in the family room. The original is painted an appropriate caboose orange, with the windows, catwalk, and the elaborate end platforms and undercarriage (lost in shadow in our photo) painted black.

All wood parts (except for the door and window frame strips) can be cut from a single sheet of ½″ plywood. Since both shaped

A. Roy DeHaan
Maple Ridge, BC, Canada

roof sections swing up, the toy-storing capacity of this box is considerable. The fixed top section simply bridges a single open box in this version, but you could place a partition under it if you wished to provide separate compartments for two children. You might even want to hinge those little center roof sections and put a bottom in that bridge, for storing smaller items. Continuous hinges are best insurance against pinched fingers.

Once you've laid out all parts, you can cut the large panel into several smaller sections so you can carry it to your table saw or bandsaw. The parts can, of course, be cut with portable power saws as well (just clamp a straightedge across the panel to guide the shoe of the saw against). Decide which cut edges will be exposed and concentrate your attention on sanding these and filling any edge voids. Then start assembly with the five panels of the basic box. Assemble the two roof sections, center them on the box to determine the position of the bridge, and glue and screw that assembly in place. Add the undercarriage assemblies and the stepped end platforms, capping these with the A panels.

You'll find it easier to paint the catwalk strips before installation—and the window areas should be painted black after you've finished painting the box orange but *before* you glue on the frame strips. If you're not handy with a lettering brush, buy white adhesive letters and numbers from a sign shop. You can always make these more durable by covering them with a clear varnish. Your rolling stock is now ready for its test run.

PANEL LAYOUT

$\frac{1}{2}'' \times 4' \times 8'$

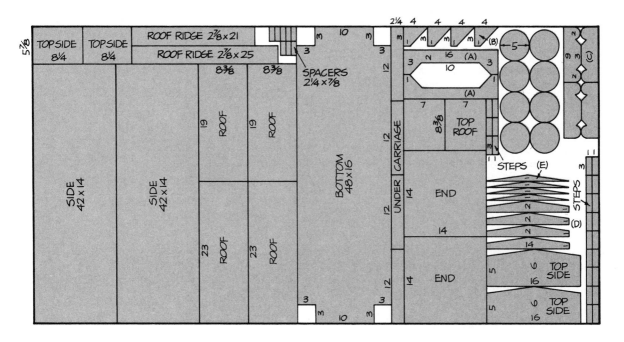

■ MATERIALS LIST ■

½"

Quantity	Description
1 4 x 8 panel	½" plywood, MDO or A-B Interior
2	½" x 16" continuous hinges with screws
2	Lid supports with screws (to prevent accidental forceful closure of lids)
Approx. 18 lin. ft.	¼" x ⅛" wood strips for door and window frames (available in hobby stores)
As required	Finishing nails, white or urea resin glue, wood dough or synthetic filler, fine sandpaper, top-quality finish

WINDOW DIMENSIONS: SIDES

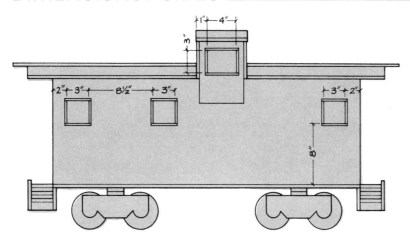

WINDOW DIMENSIONS: ENDS

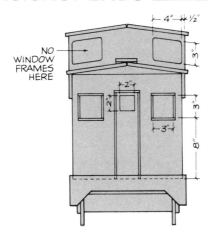

EXPLODED VIEW

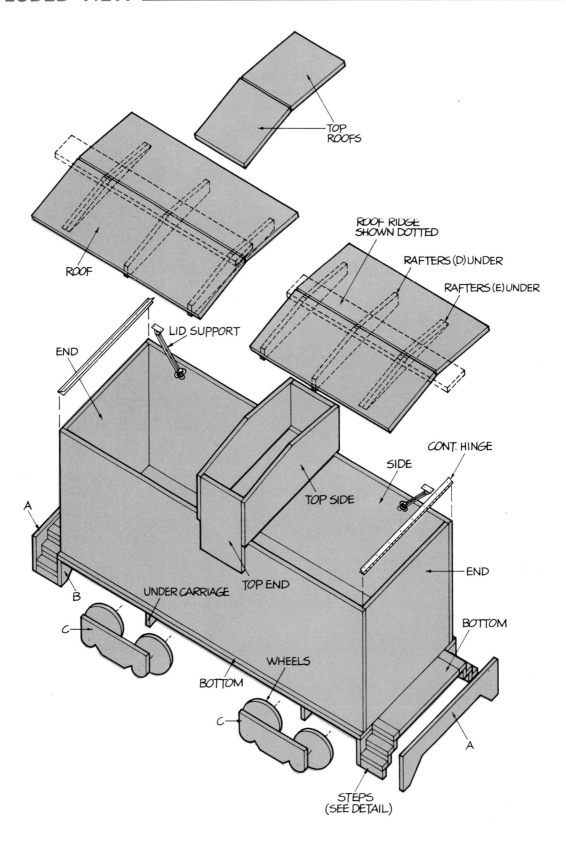

TOP
ROOFS

ROOF RIDGE
SHOWN DOTTED

RAFTERS (D) UNDER

RAFTERS (E) UNDER

ROOF

LID SUPPORT

END

CONT. HINGE

SIDE

TOP SIDE

A

TOP END

END

B

UNDER CARRIAGE

BOTTOM

C

WHEELS

BOTTOM

A

C

STEPS
(SEE DETAIL)

UNDER CARRIAGE ASSEMBLY

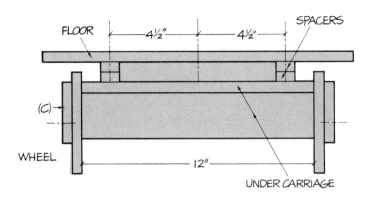

FLOOR

4½" 4½"

SPACERS

(C)

WHEEL

12"

UNDER CARRIAGE

WHEELS ASSEMBLY

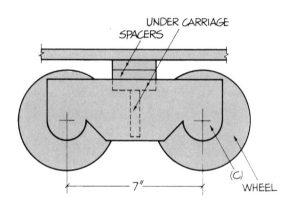

UNDER CARRIAGE

SPACERS

(C)

WHEEL

7"

ROOF ASSEMBLY

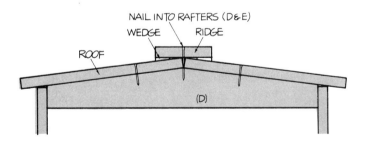

NAIL INTO RAFTERS (D&E)

WEDGE RIDGE

ROOF

(D)

STEPS ASSEMBLY

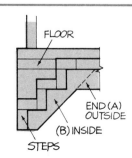

FLOOR

END (A)
OUTSIDE

(B) INSIDE

STEPS

TRUCK TOY BOX

What could bring more delight than a toybox that's a toy itself? The bigger stuffed animals can drive, because the cab top swings up for access to the front compartment. And the whole double-compartment truck rolls away on dowel axles to clear up playtime clutter. Painted dark red with some automotive striping, the toy box is so handsome you may be reluctant to garage it.

All dimensions given are finished sizes, so leave space for saw kerfs when making your layout on the ¾″ plywood panel. There's enough waste to allow for this, and to let you make the bevel cuts called for by setting your sabre saw to 15 degrees as you cut these edges.

To simplify finishing, paint the fenders, headlights, bumper,

Steven Snavely
Tiffin, OH

wheels and wheel covers—and, of course, the trim pieces—before assembly. You can touch up nail and screw heads later. The fuel tanks shown in the photo are optional. They were cut from a cardboard mailing tube, the ends plugged with turned wood blocks and the whole assembly coated with aluminum paint. Plexiglas windows (with the optional trim) add class and make the front compartment more functional. (Favored stuffed friends can be stored dust-free with the top down.) As with any toy box, it's a good idea to attach a slide to the hinged top to prevent accidental falls on small fingers.

PANEL LAYOUT

¾" × 4' × 8'

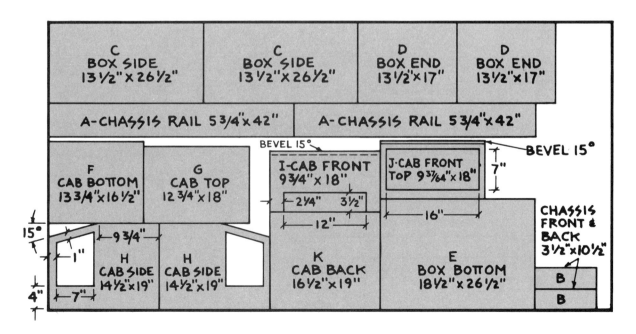

MATERIALS LIST

$\frac{3}{4}''$

Quantity	Description
1 4 x 8 panel	¾" plywood, A-B or A-D Interior or MDO
1	Dowel, ⅞" dia. x 36". Cut front axle 17½" long, rear axle 17" long.
2 pieces	1 x 1 lumber, 25⅜" long for chassis side rails
1	2 x 10 board, 6' long. Cut 6 wheels and 2 front wheel spacers.
1 piece	Scrap plywood ¼" thick x 12" x 12" for wheel covers
1 piece	2 x 4 board, 18" long for bumper
1 piece	1 x 10 board, 36" long. Cut fenders, head-lights, and grill.
10 pieces	1 x 1's, 9½" long for box vertical trim
4 pieces	1 x 1's, 2 approx. 24½" long (cut to fit) for side horizontal trim; 2 approx. 16½" long (cut to fit) for front and back horizontal trim
4 pieces	2 x 2's, approximately 13½" long (cut to fit) for outside corner molding on box (cut away ¾" x ¾" corner in each)
8 pieces	1 x 2 board, mitered 45° as shown in exploded view, for box bottom and top molding (4 cut 28" long, 4 cut 20" long)
1	Continuous hinge, 18" long for cab top
1 piece	Trim, ¼" thick x ¾" by 48" long (cut to fit) for 3 sides of cab top
2 pieces (optional)	Plexiglas (or equivalent) ⅛" x 8" x 10" for side windows. Cut to dimensions.
1 piece (optional)	Plexiglas (or equivalent) ⅛" x 8" x 17" for windshield. Cut to dimensions.
10 lin. ft (optional)	Outside corner molding, ¾", for framing win-dows. Cut to dimensions shown in detail. Cut lengths to size for windows.
170 (approx)	Wood screws, flathead, #8 x 1½" long for glue-screw assembly
As required	2d finishing nails for glue-nailing trim and molding
20	Sheet metal screws, pan head, #8 x ½" for attaching wheel covers
4	Lag screws, hex head, ¼" dia. x 1½" long for attaching wheels to axles
As required	White glue for glue-screw assembly of all joints; fine sandpaper for smoothing plywood cut edges; wood dough (if desired) for filling any edge voids; finishing materials

SIDE VIEW

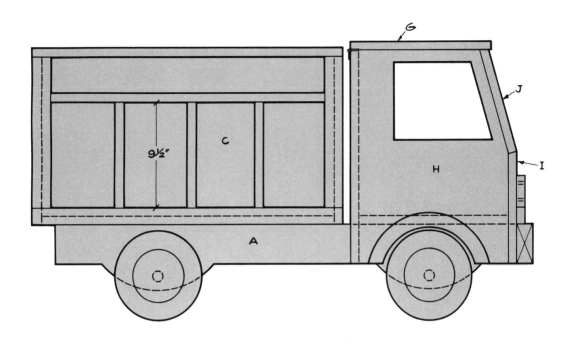

FRONT VIEW

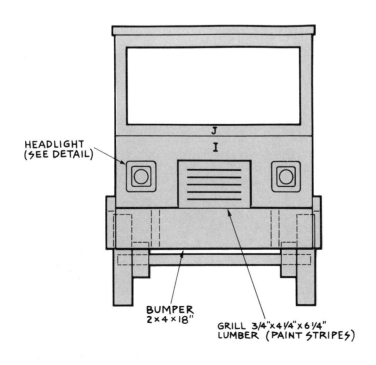

CHASSIS DETAILS

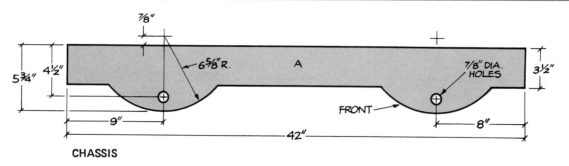

CHASSIS

CHASSIS & WHEELS: EXPLODED VIEW

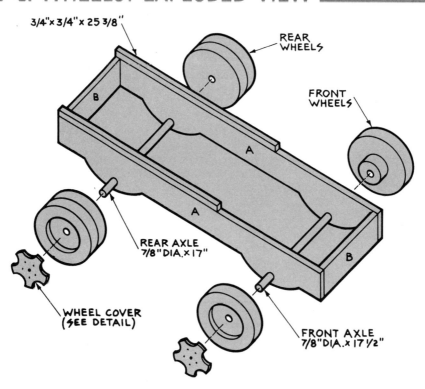

3/4"x 3/4"x 25 3/8"

REAR WHEELS

FRONT WHEELS

REAR AXLE 7/8"DIA.×17"

WHEEL COVER (SEE DETAIL)

FRONT AXLE 7/8"DIA.×17 1/2"

WHEEL COVERS DETAIL

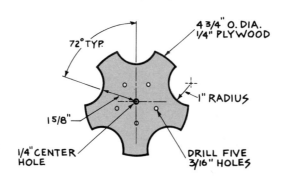

72° TYP.

4 3/4" O. DIA. 1/4" PLYWOOD

1" RADIUS

1 5/8"

1/4"CENTER HOLE

DRILL FIVE 3/16" HOLES

CAB: EXPLODED DETAIL

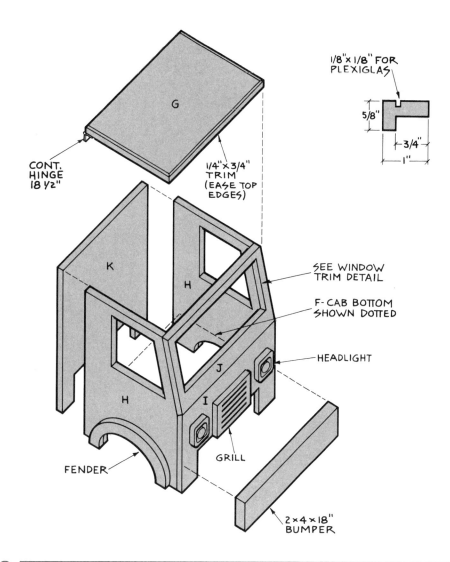

1/8"×1/8" FOR
PLEXIGLAS

5/8"

3/4"

1"

CONT.
HINGE
18 1/2"

1/4"×3/4"
TRIM
(EASE TOP
EDGES)

SEE WINDOW
TRIM DETAIL

F-CAB BOTTOM
SHOWN DOTTED

HEADLIGHT

FENDER

GRILL

2×4×18"
BUMPER

CAB DETAILS

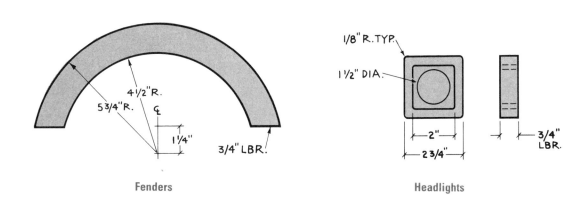

4 1/2" R.

5 3/4" R.

1 1/4"

3/4" LBR.

Fenders

1/8" R. TYP.

1 1/2" DIA.

2"

2 3/4"

3/4"
LBR.

Headlights

TOY BOX: EXPLODED VIEW

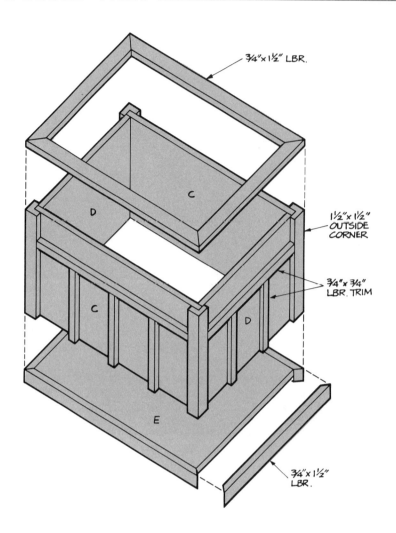

¾"x1½" LBR.

1½"x1½" OUTSIDE CORNER

¾"x¾" LBR. TRIM

¾"x1½" LBR.

WHEELS DETAILS

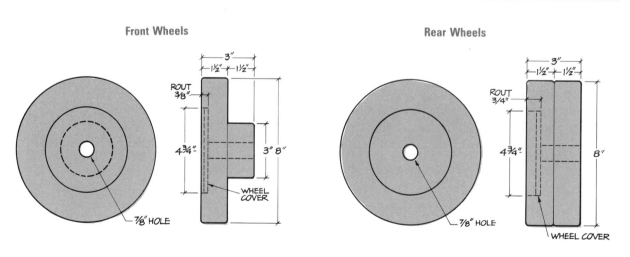

Front Wheels

3"
1½" 1½"
ROUT ⅜"
4¾"
3" 8"
WHEEL COVER
⅞" HOLE

Rear Wheels

3"
1½" 1½"
ROUT ¾"
4¾"
8"
⅞" HOLE
WHEEL COVER

THREE-POSITION NURSERY WHALE

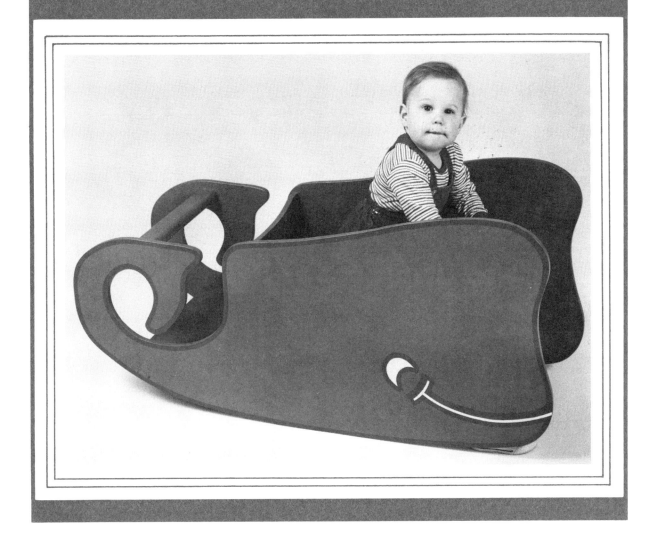

On its tummy, this whale is a gentle rocker (as demonstrated above); on its back, it's a sturdy play desk; upended, it's a highchair (see photo next page), but add a safety belt for active diners. Best of all, for the nursery set, it's a whale they can "steer"— there's a just-pretend wheel of plywood disks that you can assemble to rotate, if you wish. (In our sketches, it's shown as a fixed handle.) When youngsters tire of riding the *swimming* whale, beach him belly-up and let them reverse position, using the wheel-bar as a bench and the now-upside-down bench as a play table, ideal for crayons or finger paints.

All this from just a half-panel of ½" plywood, a few nails and screws, glue and paint. You'll need a sabre saw (or heavy duty

Jamie Macouzet
Peubla, Mexica

jigsaw) to cut the whale shapes, after you've enlarged our graph-squared pattern by laying out a 2″ grid on the back surface of your plywood (it's then quite easy to place the right amount of any curve in each square it crosses). The only other tools you'll need are hammer, screwdriver, drill and a couple of pipe clamps—although you can do without those if you rig up your own tourniquet clamp with rope.

Carefully examine all cut edges for any voids, and fill them with wood dough or small glue-blocks. Sand all edges smooth and apply an overall prime coat. We finished our whale in a bright blue, then accented all edges in a darker blue, carrying it over a half inch onto the surfaces for a softening effect. To assure the smoothest paint surfaces and splinter-free edges, MDO (medium-density overlaid) plywood is strongly recommended for nursery furniture.

As you can see in our photo right , the sturdy inch-thick tail brace does serve as a restraint bar when the whale stands on its nose to serve as a high-chair. But an active child of the age of our model can squirm under the bar and drop to the floor. That's why we proposed the safety belt—any 2″-wide strip of soft but strong fabric screwed to the inside of the whale's body at one end so it can be anchored with a sturdy snap at the other. This lap restraint also dampens the tendency of youngsters to rock a high chair.

It's worth the extra effort to add the third function to this project. After all, how many chances will you ever get to invite a whale to your dinner table?

PANEL LAYOUT

½″ × **4′** × **4′**

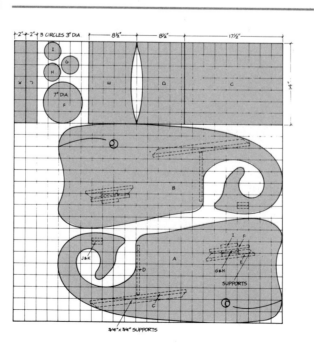

MATERIALS LIST

Quantity	Description
1 4 x 4 panel	½″ plywood, A-B Interior, A-B or A-C Exterior, or MDO
5 ft	¾″ x ¾″ lumber for seat supports
30	#8 screws, 1½″ long for installing Parts C, D, E, and assembly J-K
6	6d casing or finishing nails for assembling steering wheel
As required	White glue for assembly
As required	Wood dough for filling any small voids in plywood cut edges; fine sandpaper for smoothing cut edges and cured wood dough
As required	Primer and paint

EXPLODED VIEW

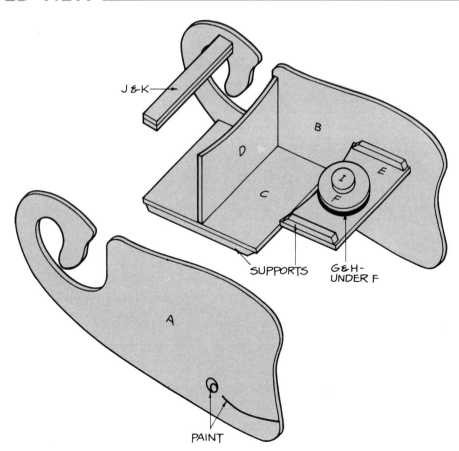

J & K

B

D

E

C

I

F

SUPPORTS

G & H-
UNDER F

A

PAINT

TOP, FRONT, & BACK VIEWS

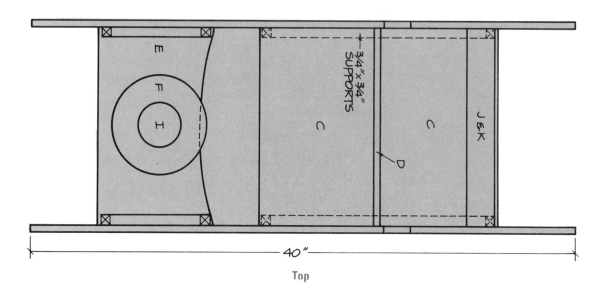

Top

Front

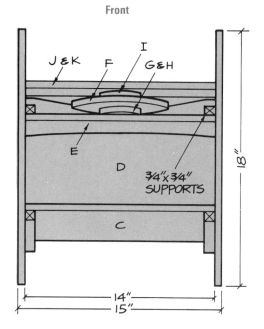

Back

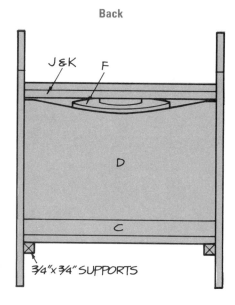

SEATS SUPPORT DETAIL

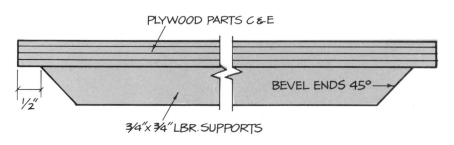

PLYWOOD PARTS C & E

BEVEL ENDS 45°

½"

¾" x ¾" LBR. SUPPORTS

LAMINATED SPIRAL STAIRCASE

A laminated sculpture (albeit a functional one), this lovely staircase was built into the contemporary home of its designer, a professor of design at a midwestern university. "I wanted a unique but relatively inexpensive wooden stair which could be prefabricated in modules in my workshop," he told me, "and then transported easily to the building site for assembly and finishing."

You'll need 56 keyhole-shaped treads and 88 disks—all cut from sheets of ¾" fir plywood—plus 16 disks cut from a half-sheet of ½" ply. Each of the 11 treads is laminated from five keyholes, to a total thickness of 3¾". Six ¾" disks and one ½" disk are laminated into cylinders 5" tall for the 11 risers. The remaining

Jerry Tow
Ames, IA

PANEL LAYOUTS

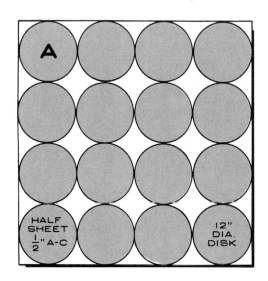

½″ × 4′ × 8′

A

HALF
SHEET
½″ A-C

12″
DIA.
DISK

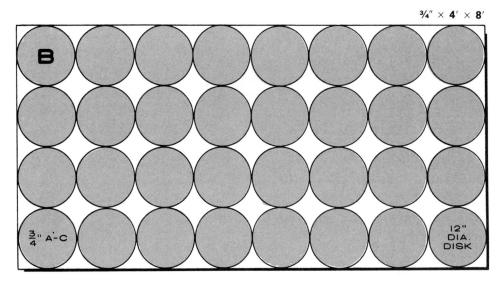

¾″ × 4′ × 8′

B

¾″ A-C

12″
DIA.
DISK

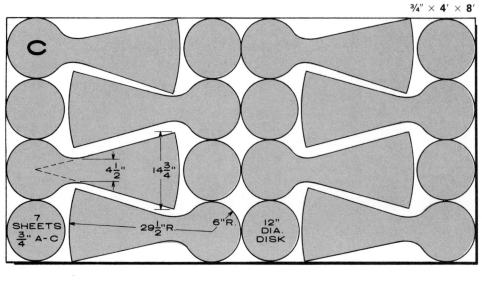

¾″ × 4′ × 8′

C

4½″

14¾″

7
SHEETS
¾″ A-C

29½″R.

6″R.

12″
DIA.
DISK

¼″ × 2′ × 8′

D

3″ WIDE STRIPS TO BE SOAKED AND BENT FOR RAILING

HALF SHEET (2′ x 8′) ¼″ A-C

MATERIALS LIST

1 $\frac{1}{2}$　　8 $\frac{3}{4}''$　　1 $\frac{1}{4}''$

Quantity	Description
8 4 x 8 panels	¾″ plywood, A-C Exterior or A-D Interior
1 4 x 4 panel	½″ plywood, similar grade
1 2 x 8 panel	¼″ plywood, A-C Exterior
1	¼″ x 12″ x 12″ steel plate
1	⅞″ x 12′ steel threaded rod
8	nuts and washers for ⅞″ bolt
50	#8 x 2″ wood screws
22	#8 x 1″ wood screws
11	hardwood posts, 1 x 1, 42″ long
As needed	Woodworker's glue, plastic wood, sandpaper (40, 80, & 120 grits), sanding sealer, satin polyurethane varnish, enamel for handrail

disks are laminated to form the top of the central column. A ⅞″ hole is drilled in the center of each riser, and the cylinder of each tread, for assembly.

The dimensions given here result in a staircase five feet in diameter by 11 feet tall, overall, fitting a subfloor-to-subfloor height of 104½″. The height of the risers is easily adjusted to other floor-to-floor dimensions, but keep all step-spacing uniform: You invite stumbles if you vary tread-to-tread distance in any stairway. Bear in mind that in most locations you'll have to add an extra floor joist beneath the central column to accommodate the additional weight.

First, cut full-size templates of the keyholes from cardboard. Trace the pattern as shown on seven of your eight sheets of ¾″ plywood, filling in with plain 12″-dia. disks, laid out with a compass. Scribe 32 additional 12″ circles on the eighth panel, then trace 16 more on the half sheet of ½″ ply. The original designer cut out all these pieces with a saber saw, but if you find that prospect tedious, you can cut full-panels into half panels, tack them temporarily together and cut two or three at once on a full-size bandsaw. Wherever possible, cut ⅛″ oversize.

When you stack your keyhole shapes for each tread-unit, select a good ("A") face for the top and bottom of each unit. (Five keyholes per step and 11 steps leaves one keyhole extra: That's for tying the staircase to the upper floor.) Hand-screws or short bar clamps are applied to each stack until the glue sets.

Next laminate the six disks of ¾″ ply plus one of ½″ ply—or

COLUMN & TREAD ASSEMBLY

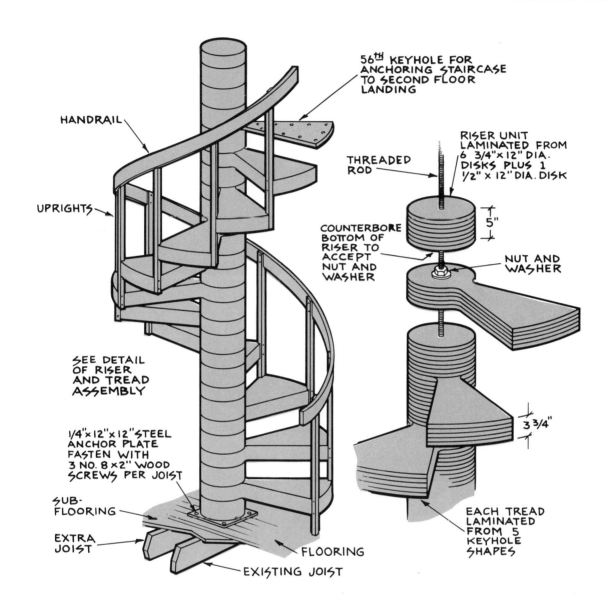

56TH KEYHOLE FOR ANCHORING STAIRCASE TO SECOND FLOOR LANDING

HANDRAIL

UPRIGHTS

THREADED ROD

RISER UNIT LAMINATED FROM 6 3/4"x12" DIA. DISKS PLUS 1 1/2" x 12" DIA. DISK

5"

COUNTERBORE BOTTOM OF RISER TO ACCEPT NUT AND WASHER

NUT AND WASHER

SEE DETAIL OF RISER AND TREAD ASSEMBLY

1/4"x12"x12"STEEL ANCHOR PLATE FASTEN WITH 3 NO. 8×2" WOOD SCREWS PER JOIST

3 3/4"

SUB-FLOORING

EXTRA JOIST

EACH TREAD LAMINATED FROM 5 KEYHOLE SHAPES

FLOORING

EXISTING JOIST

RAILING CLAMP

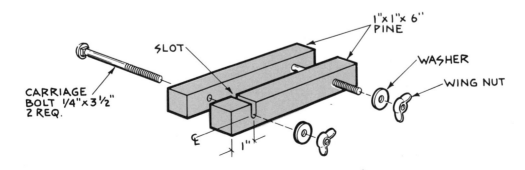

1"x1"x 6" PINE

SLOT

WASHER

WING NUT

CARRIAGE BOLT 1/4"x3 1/2" 2 REQ.

1"

whatever other combination you've determined will fill your floor-to-floor space—until you've assembled 11 riser units. Laminate 21 of the remaining 22 disks of ¾" ply with four of the ½"-ply disks to make four modules for the top of the center post. (If you want a center post two feet higher at the top than the one shown, you'd have to cut 32 more disks from another full sheet of ¾" ply and continue the lamination.)

When the glue has set, refine the shapes of each step and riser cylinder by sanding with a belt sander (or, in the case of the risers, by truing up on a wood lathe). Drill a ⅞" hole accurately in the center of each cylinder.

To assemble, drill a ⅞" hole in the center of a ¼" steel plate, one foot square. Insert one end of a 12' ready-bolt and secure the plate with a washer and nut on each side. Counterbore your subfloor to accept the bottom nut, and fasten the steel plate to the subfloor with six #8 × 2" wood screws, with three of the screws going into each joist (an original joist and the extra joist you installed).

It's a good idea to drill a ⅞" hole in a scrap 2 × 4 and slip it over the top of the ready-bolt to stabilize it during assembly. Starting from the top of the bolt, slip on one 5"-thick riser cylinder and slide it all the way to the bottom. Coat the top of the cylinder with glue and slide a tread unit down on top of it—taking care to position this first step for the correct beginning of the spiral.

RAILING ASSEMBLY

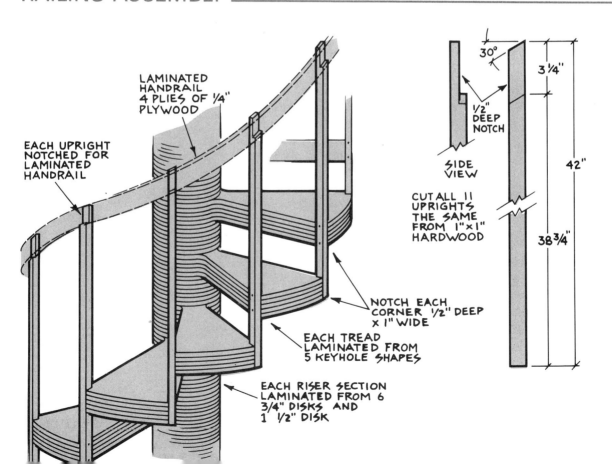

Now continue gluing and adding risers and treads, aligning the leading edge of each new tread with the trailing edge of the previous tread.

After every other step, turn on a washer and a ⅞" nut. Snug down tightly and counterbore the bottom of the next riser column. Even so, you'll probably want to prop up the outer edge of each tread unit to assure it remains level while the glue sets.

When the last tread is in place, add the left-over keyhole piece at the correct angle to tie the staircase to the upper floor. Then add as many disks as desired to raise the center post. Put on the last washer and nut and tighten, then trim the ready-bolt flush with the top of the nut. Counterbore the bottom of a ¾" circle and glue it to the top of the post and cap this with the final ½" disk.

Fill any voids in the exposed plies with wood dough. When this patching sets, sand all edges, starting with 40 grit and finishing with 120.

For the railing, cut 11 uprights 42" long (the original prototype used full 4/4 oak, but there are a number of woods that will compliment the laminated ply). Notch the top of each upright ½" deep by 3¼" wide on the outside surface, to accept the handrail, cutting both the bottom of the notch and the top of the upright itself at a 30-degree angle. To attach the uprights, notch each tread at both outer corners as shown. Each upright should fit flush with the curved edge of the tread, ½" into the trailing edge of the lower step and ½" into the leading edge of the higher step. Check verticality with a level and attach with two #8 × 2" wood screws into each step.

For the handrail, rip eight strips from the vertical half sheet of ¼" plywood. Soften the strips one at a time, with steam or hot water (boiling water applied with a paint brush is often sufficient). Bend and attach the strips to the notches in the uprights with C-clamps, placing the "A" side of the ply so it faces the staircase. Let the lower end of the handrail extend 2½" beyond the first upright.

Continue to wet, bend and clamp the remaining strips until the handrail is four plies thick, removing and reapplying the C-clamps as you go. Be sure the "A" faces of the final strips face out. You'll need many clamps, so you may want to make up a couple dozen of them out of 1" × 1" pine and carriage bolts, as shown in a sketch.

When all strips have dried thoroughly, disassemble carefully and reassemble by face-gluing, clamping and attaching to each upright with two #8 × 1" screws. Attach the upper end of the handrail (which extends beyond the top upright) to the wall or railing at the top of the staircase.

After entire staircase and railing has been sanded, apply sanding sealer. Resand lightly, then apply two or three coats of polyurethane varnish, sanding lightly between each coat. (In the prototype, the handrail is enamelled gloss black.)

OUTDOOR PROJECTS

FOUR KNOCKDOWN CHAIRS

A set of four high-style lawn chairs is what you get from a single ¾"-thick sheet of plywood—with nothing wasted but the holes. (And if you cut these out carefully with a scroll saw, you'll have a bonus set of discus for eight throwers.)

We chose this top-prize-winning project to kick off our selection of outdoor projects because of its appealing simplicity. Every home should have a quartet of these "extra" chairs—the seats and backs simply slot into the sides; each chair is as quick to set up as a folding bridge chair, and it stores in even less space. Though sturdy and comfortable (especially if you add a seat cushion), each chair is nothing more than the three flat panels—there's no construction involved: no nailing, gluing or bracing

Charles T. Goulding
Elkins Park, PA

PANEL LAYOUT

¾" × **4'** × **8'**

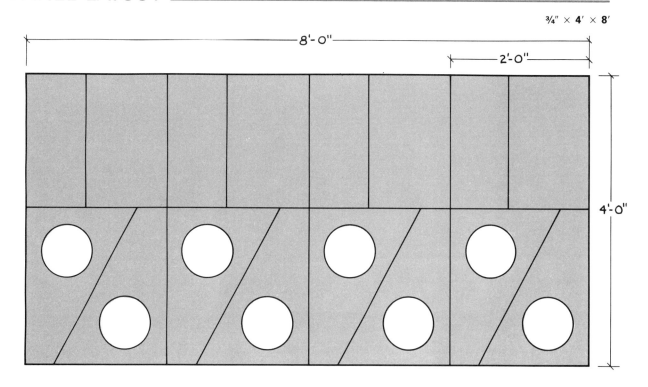

CHAIR DIMENSIONS

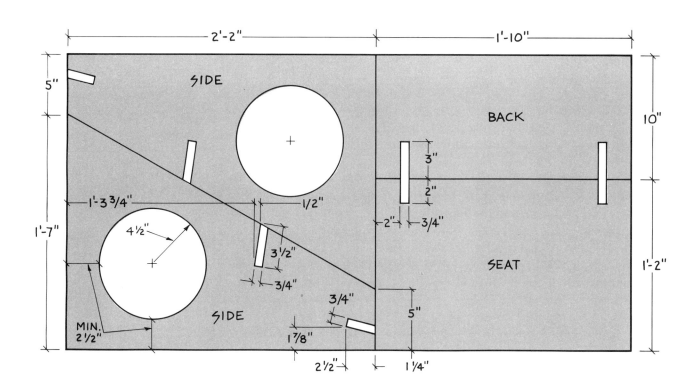

of any kind. And they're just as handy for extra guests *indoors* as out, though we find them ideal for deck or patio.

Layout couldn't be simpler; only the slot locations are critical. Get a sheet of wrapping paper measuring 2′ × 4′ and do the layout for a single chair, below. Transfer this to your plywood

■ MATERIALS LIST ■

Quantity	Description
1 4 x 8 panel	¾″ plywood, MDO, or A-B or A-C Exterior
As required	Wood dough or other filler for filling cut-edge voids, if desired
As required	Fine sandpaper for smoothing plywood cut and filled edges; exterior primer and paint
1 qt. ea.	Two colors of good outdoor enamel, plus undercoat (chairs shown are two-tone)

sheet, repeating it three more times. Remember that in the basic layout no provision is made for the width of saw kerfs, so keep your blade centered on the mark. This doesn't apply, of course when it comes to cutting the slots. We made our circular cutouts 9″ across, but note that (for strength) they should never come closer than 2½″ to the plywood edges. The beauty of so many identical parts is that you can stack them and cut several slots and holes at once.

We chose MDO plywood for our chairs, and kept the slots a snug slip fit. Remember, you'll increase the panel thickness a bit when you add a prime and finish coats of enamel. We finished our set in two colors (butter yellow and Chinese red), and when we set them up each time, we give ourselves the option of mixing or matching the panels. It adds variety to our entertaining. And the stacked panels of alternating colors are even fun to look at in storage.

Before painting, however, be sure to examine all cut edges for exposed voids in the inner plies. You'll want to patch these with wood dough and sand all edges smooth, rounding corners slightly, especially on the seat panels.

FRONT VIEW

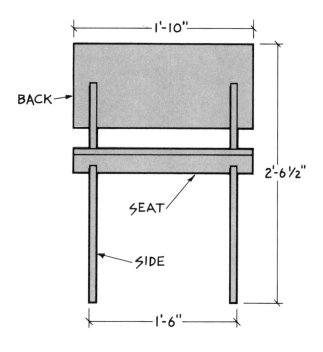

BACK

1'-10"

2'-6½"

SEAT

SIDE

1'-6"

SIDE VIEW

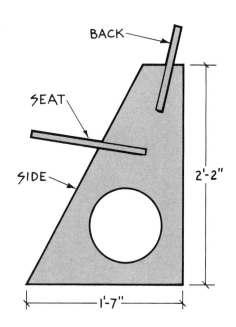

BACK

SEAT

SIDE

2'-2"

1'-7"

EXPLODED VIEW

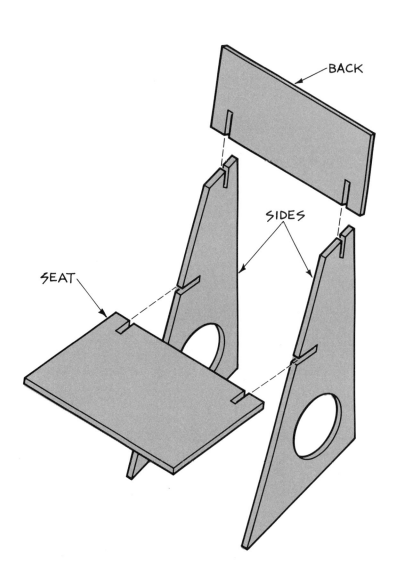

BACK

SIDES

SEAT

PAIR OF LAWN CHAIRS

Kerf-bent arms of this matched pair—both, incredibly enough, cut from a half sheet of A-C ⅝″—demonstrate plywood's versatility. After the seat and back panels are cut from the center of each quarter sheet, the remaining frame is repeatedly kerfed, just short of ½″ deep, on its back face. Strips of masking tape are applied opposite each cut, on the "A" face, the kerfs are filled with glue and the arms are bent until the kerfs close.

The unique styling (we painted our pair bright orange) make these chairs appropriate for backyard entertaining or relaxing by a pool. Keep them finished with a waterproof enamel and they'll last for years.

You can do the main "separation" cuts with a portable circular

Gary A. Kerr
San Diego, CA

245

saw, stopping short of all interior corners. Complete these round corners with a saber saw, as shown in the panel layout, then cut the seat notches. Set the blade at 45 degrees and cut the backrest notches. Bevel the bottom edge of the back to butt against the seat.

Space the bending kerfs 1⅛" apart, as shown. The simplest way to achieve the bending is to cut templates of scrap plywood to the dimensions shown in the detail. It's a good idea to wax the edge grain of the templates to prevent adhesion of any squeezed-out glue. Position a template at each side of the arm panel and slowly pull the two panel ends toward one another. Tack the templates to the now-curved arms as shown, to clamp them in position while the glue sets. You can also glue-nail the notched seat pieces in place at this time. (You'll need at least one helper for the bending operation.)

After the glue sets, carefully withdraw the clamp-nails and fill the holes they leave with wood dough. Check the underside kerfs and fill any of these that need it. Fill any edge voids, while you're at it. Screw the chair back to the seat and nail the backrest in place with the 2d finishing nails.

After thoroughly sanding all edges and rounding corners slightly, apply a prime coat followed by two finish coats of a good outdoor enamel.

PANEL LAYOUT

⅝" × 4' × 4'

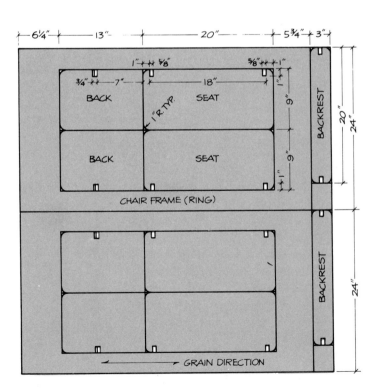

Actually, if done properly, kerf bending results in units of exceptional strength. But if you have any doubts, invest in a second half-sheet of good plywood (or buy a full sheet and use the other half) to substitute for the scrap templates. In this case, of course, you don't wax the edge grain—you apply glue before bending the arm panel around—and you drive finishing nails all the way in (and leave them in place). The chairs will still be handsome with the closed arms, though they'll be quite a bit heavier. You might want to paint these closure panels a second color, for a two-tone effect.

MATERIALS LIST

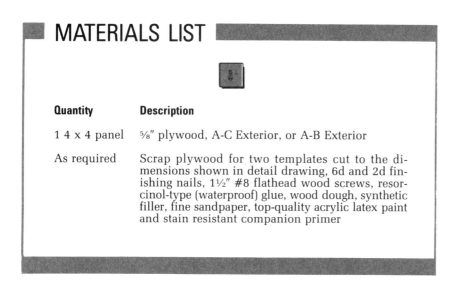

Quantity	Description
1 4 x 4 panel	⅝" plywood, A-C Exterior, or A-B Exterior
As required	Scrap plywood for two templates cut to the dimensions shown in detail drawing, 6d and 2d finishing nails, 1½" #8 flathead wood screws, resorcinol-type (waterproof) glue, wood dough, synthetic filler, fine sandpaper, top-quality acrylic latex paint and stain resistant companion primer

ARM DETAIL

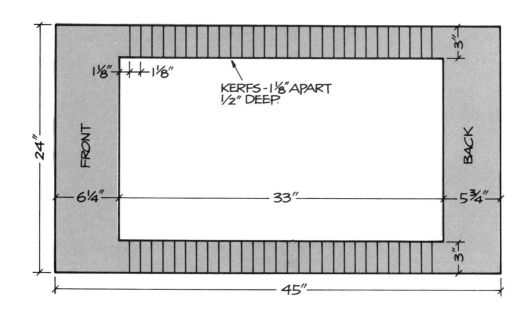

TEMPLATE DIMENSIONS

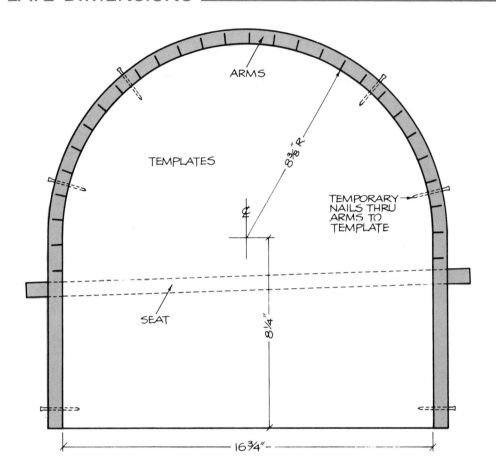

ARMS

TEMPLATES

8⅜" R

C

TEMPORARY
NAILS THRU
ARMS TO
TEMPLATE

SEAT

8¼"

16¾"

ISOMETRIC VIEW

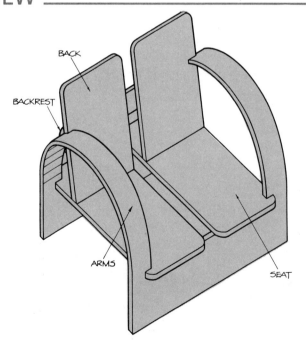

BACK

BACKREST

ARMS

SEAT

FRONT, SIDE, & TOP VIEWS

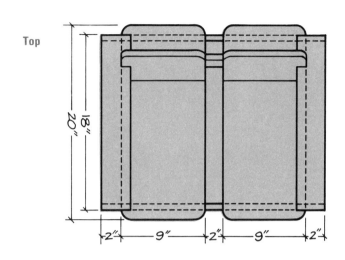

Top

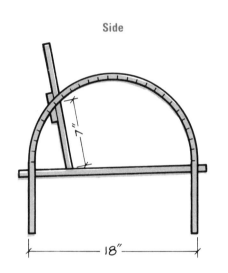

Side

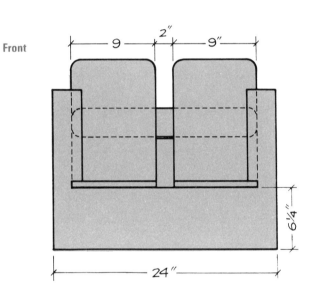

Front

BACKREST DETAILS

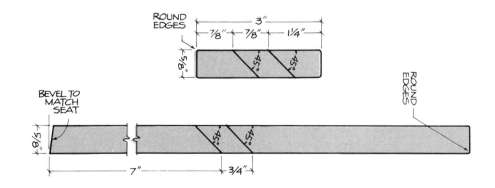

DRUM SLING CHAIR

Two drum-like cylinders of taut rope support the canvas sling, avoiding the discomfort of cross rods at your knees and head. As you sit, these flexible supports adjust the sling to your contours, softly cradling your weight. Perfect for deck, patio or poolside, the unit is slotted together for quick disassembly at the end of the outdoor season.

You'll need a steel square and a compass to lay the parts out on the panel as shown on the next page. A saber saw with a clamped straightedge is the best bet for cutting out these intricately-nested parts. The four 6"-dia. disks are laid out for drilling as in the detail. The ⅜" holes for the ropes are spaced with their centers ½" from the edge. To cut down on chip-out while drilling,

Ken Kramer
Carbondale, IL

you may want to drill all these holes before cutting out the disks. Indeed, if you have a drill press, it would be best to cut rough blanks for the four disks, stack them and tack them temporarily together, drill the holes through all four at once, then carry the stack to a bandsaw for final shaping. Before splitting the stack, drill the ⅜″ hole through dead-center for the threaded rods. Now notch two of these disks as shown in the detail, so they'll rest against the top cross-piece.

You could handle the 4″-dia.-disks much the same, though the only drilling they require is the ⅜″ center hole. Glue one small disk to each large disk, aligning their center holes.

Complete shaping of all frame members, drilling through-holes for the two rods, and socket-holes for the two dowels. The latter are to secure the canvas sling at each end. Note that the bottom end is held by a fixed dowel, but the upper end is sewn around a dowel that can be positioned in several pairs of holes, to stretch the canvas tauter or let it go slack, as an adjustment to your own comfort. To reposition this dowel, you just pull up the top cross-piece to free its slats; this lets you spread the two side members enough to free the dowel. Replacing the cross-piece locks the dowel in its new position.

PANEL LAYOUT

½″ × **4**′ × **4**′

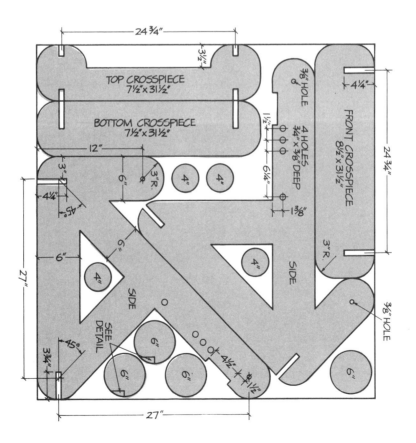

Cut the threaded rods just long enough to project ¾" on each side of the assembly. Mark them at the inner face of each small disk and turn a nut onto each end so that a washer will sit flush with this mark. Insert the rods through the disks and lock the disk assemblies in place with another washer and nut. Now you're ready to thread the rope back and forth between aligned holes, as shown.

▪ MATERIALS LIST ▪

$\frac{1}{2}$

Quantity	Description
1 4 x 4 panel	½" plywood, MDO or, if available, A-C Exterior
53 lin. in.	⅜" dia. threaded steel rod with 8 nuts and washers
46 lin. in.	¾" dia. hardwood dowel
50 lin. ft	¼" vinyl-covered twisted wire clothesline, or woven cotton clothesline
4	1¼", #10 flathead wood screws
2 yds.	Canvas cloth
As required	Fine sandpaper, white or urea resin glue, top-quality enamel paint and companion primer

DRUM DETAILS

Rope Assembly

ROPE

Assembly

Dimensions

⅜" x 26½" THREADED ROD

15°

¾"

4" DIA.

6" DIA.

⅜" HOLES FOR ROPES

⅜" HOLE FOR ROD

90°

THIS PIECE CUT OUT OF TWO 6" CIRCLES – (TO REST AGAINST TOP CROSSPIECE)

½"

WASHERS & NUTS

EXPLODED VIEW

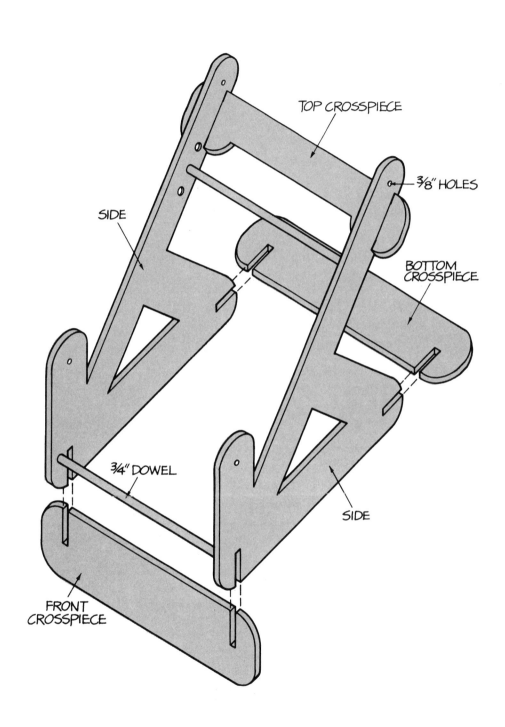

TOP CROSSPIECE

⅜" HOLES

SIDE

BOTTOM CROSSPIECE

¾" DOWEL

SIDE

FRONT CROSSPIECE

PORCH/LAWN GLIDER

Take yourself back to the good ol' days—in miniature. Remember those sunny summer afternoons, an icy glass of lemonade and a carefree hour reading or chatting in the lawn swing? Here's a compact version (cut completely from a single sheet of ¾" plywood and assembled with a few pieces of hardware) that's so charming you may want to move it into the family room at season's end, and toss a nice upholstered seat cushion on it. It's a spacious single seater or a cozy bench for two.

A simple double-pivot linkage (see final detail) hangs the bench portion inside the base for a gentle swinging action. Just be sure to use galvanized hardware—and exterior-grade plywood. To prevent chipping the face veneers (since both sides of most pieces

Paul Senecal
Badger, CA

are exposed) outline all slots with a sharp utility knife and metal ruler before using your saber saw. Another tip: Before you drill holes in the spacers, drill the legs and use them as a template. When all parts are ready for assembly, sand all edges, rounding corners enough to make the unit "user friendly." Fill any edge voids with wood dough and, when it sets, give a final sanding with fine grit paper.

We enameled our prototype in two colors, painting the seat section a buttery yellow and the base support a milk-chocolate— tasty colors for a bright green lawn.

Study the main drawing and the front and side elevations to see the unusual way this unit slots and bolts together. Most of the nuts are centered in 1" holes drilled at the base of ¼" holes *edge*-drilled to take ¼" bolts. You hold the nut in the hole while you turn the bolt snugly into it. It makes for a sturdy assembly without the need for glue and cleats, and gives you the option of taking the unit apart for storage or transport. Only the top spacers and the arm rests are permanently attached with glue and finishing nails.

If you can't find galvanized flat iron, you can keep the linkage from rusting by painting it to match the base. Just buy and apply a metal primer that's compatible with the top coat you've chosen. A drop of heavy-duty oil on each pivot at the beginning of the summer should keep you swinging peacefully for the season.

PANEL LAYOUT

¾" × 4' × 8'

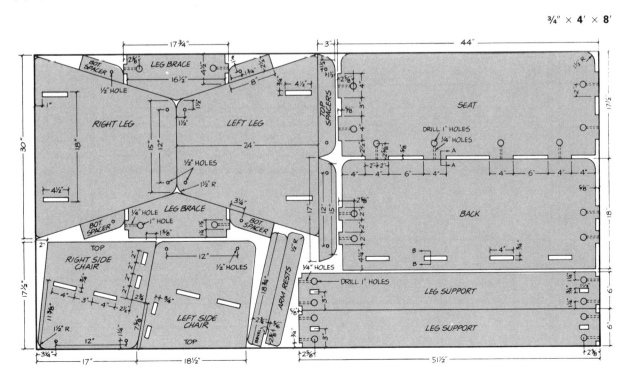

▇ MATERIALS LIST ▇▇▇▇▇▇

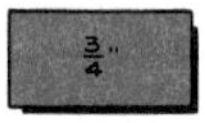

Quantity	Description
1 4 x 8 panel	¾″ plywood, MDO, or A-B or A-C Exterior
8	½″ x 3″ zinc-plated hexagonal head bolts with 24 washers and 8 nuts, bolts drilled at ends for cotter pins (see seat hinge detail)
8	Cotter pins (see seat hinge detail)
4	⅛″ x 1″ x 16″ flat iron (see seat hinge detail)
24	¼″ x 3″ zinc-plated hexagonal head bolts with 24 ¼″ x 1¼″ and 24 ¼″ x ¾″ washers, and 24 ¼″ square nuts (see attach system detail)
6	1½″ flathead wood screws (for arm rests)
As required	White or urea resin glue, wood dough or synthetic filler, fine sandpaper, top quality finish.

FRONT VIEW

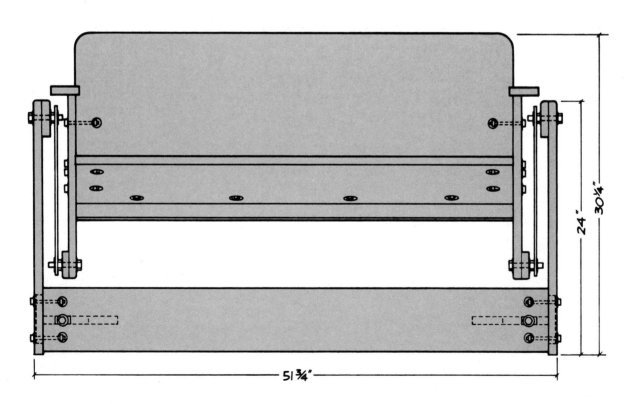

SIDE VIEW

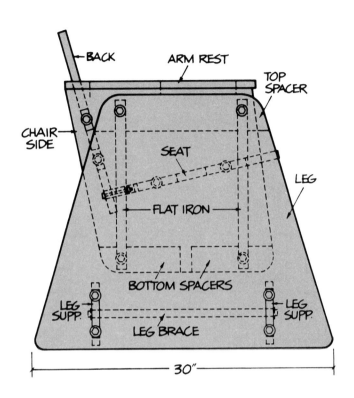

BACK

ARM REST

TOP SPACER

CHAIR SIDE

SEAT

LEG

FLAT IRON

BOTTOM SPACERS

LEG SUPP.

LEG SUPP.

LEG BRACE

30"

DETAILS

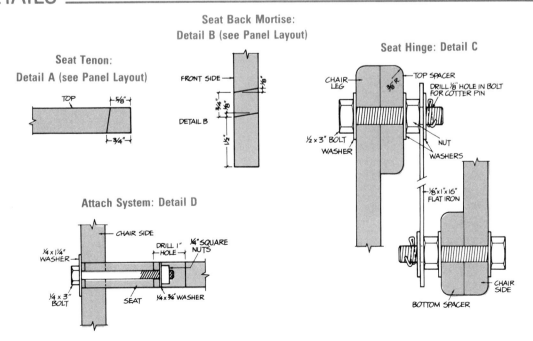

Seat Back Mortise:
Detail B (see Panel Layout)

Seat Tenon:
Detail A (see Panel Layout)

TOP
5/8"
3/4"

FRONT SIDE
1/8"
3/4"
1/8"
1/2"
DETAIL B

Seat Hinge: Detail C

CHAIR LEG
3/8 R
TOP SPACER
DRILL 1/8" HOLE IN BOLT FOR COTTER PIN

1/2 x 3" BOLT
WASHER
NUT
WASHERS

1/8" x 1" x 16" FLAT IRON

CHAIR SIDE

BOTTOM SPACER

Attach System: Detail D

CHAIR SIDE
1/4 x 1 1/4" WASHER
DRILL 1" HOLE
1/4" SQUARE NUTS
1/4 x 3" BOLT
SEAT
1/4 x 3/4" WASHER

PATIO DINING SET

As the contest judge, I had to admire the clever way this entry utilized its two sheets of ¾″ plywood to get a 40″-square table and four adult chairs! All those stylish half-circle cutouts in the chair sides would have been waste on the cutting diagram, so the designer shrewdly cut each one in half to form a set of four quarter-round seat braces per chair. And when the layout was complete, there was still enough left over to create a handy 20″ two-tier lazy susan tray for the center of the table.

Just for the fun of it, start with this first. Buy a 6″ bearing and mount it between the 20″ disk and its 9″ base. Drill holes for a ⅜″ dowel through the three small disks, and drill socket holes ⅜″ deep in the top face of the 20″ disk and the bottom face of the 9″ top. Assemble this five-disk stack with glue.

Dan Knapp
Sedalia, MO

258

The table assembly isn't really complex—it's just a matter of butting strips together and fastening with glue and finishing nails. Start with the four leg L's; then apply the edge-strips to the table top to form a shallow tray. Secure a leg assembly at each corner using glue and screws driven from inside the legs. For a sturdier

PANEL LAYOUTS

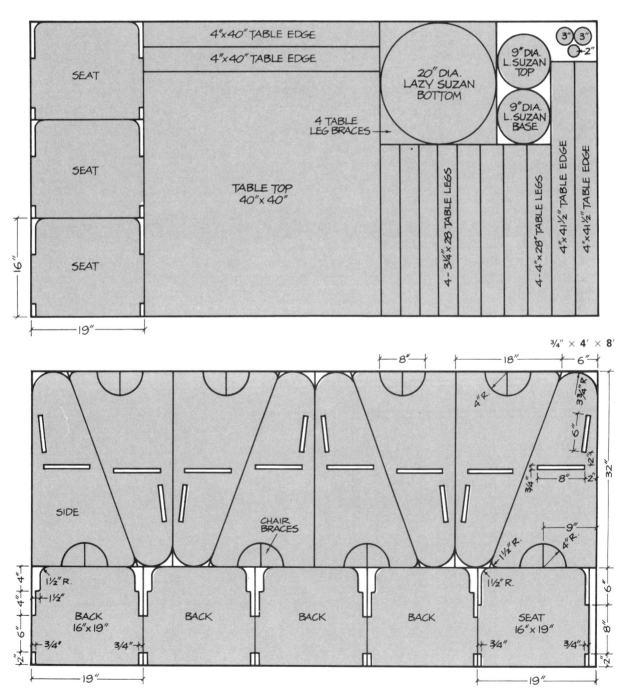

■ MATERIALS LIST ■

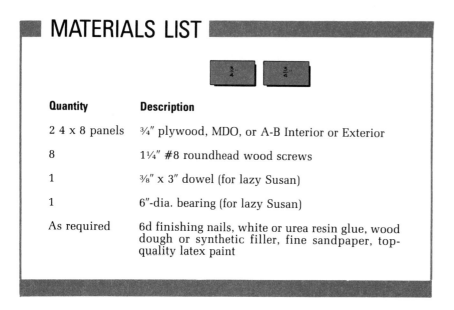

Quantity	Description
2 4 x 8 panels	¾″ plywood, MDO, or A-B Interior or Exterior
8	1¼″ #8 roundhead wood screws
1	⅜″ x 3″ dowel (for lazy Susan)
1	6″-dia. bearing (for lazy Susan)
As required	6d finishing nails, white or urea resin glue, wood dough or synthetic filler, fine sandpaper, top-quality latex paint

assembly, you may want to add metal leg braces. They won't show when you invert the table.

Assembly of the four chairs couldn't be simpler—they just slot together. But assembly is permanent, with glue on the tabs. We filled the cracks with wood dough, sanded the outside surfaces smooth and applied two coats of a good exterior enamel. Screws driven through sides and seat to secure the braces had been countersunk, filled, and sanded, so the paint concealed all.

SEAT VIEWS

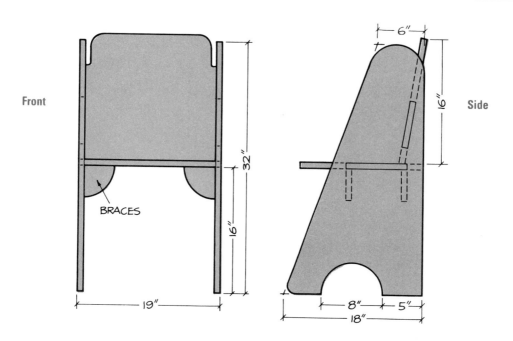

TABLE: BOTTOM VIEW

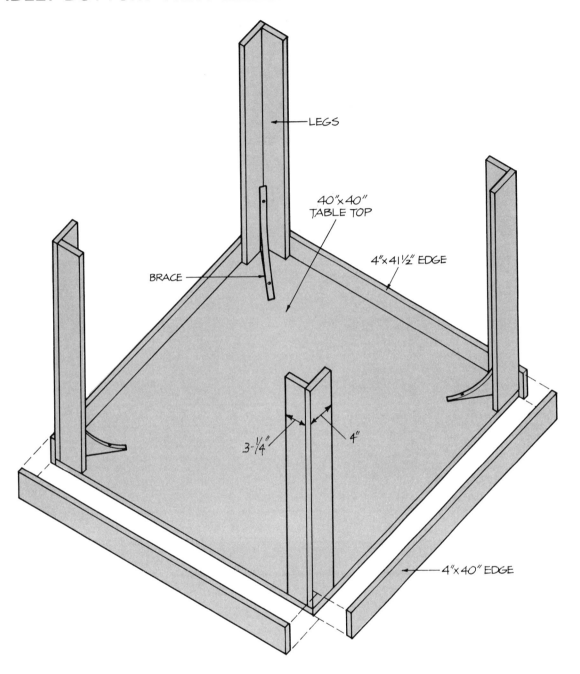

LEGS

40"x 40" TABLE TOP

4"x 41½" EDGE

BRACE

3-¼"   4"

4"x 40" EDGE

LAZY SUSAN ASSEMBLY

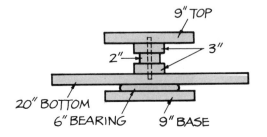

9" TOP

3"

2"

20" BOTTOM
6" BEARING   9" BASE

BUTCHER BLOCK TABLE

These butcher-block surfaces are strictly acquired—this is a plywood project that was veneered with plastic laminate in a wood-strip pattern. It's appropriate, since the design is based on a traditional butcher's carving table, but if you don't fancy applying laminates to all those surfaces, you can simply stain the plywood surfaces to an oak or maple tone—or use MDO plywood and paint the unit in one or more colors.

Whatever your finish, you end up with a sturdy table that provides an ideal height for stand-up food preparation, with lower-level drop-leaves that are just right for informal dining for two (don't forget the hurricane lantern). Finally, at mid-level there's a breadboard that pulls out of a slot. Since it stores beneath the

Arthur R. Burns
Los Angeles, CA

table's top, it's always clean, and it extends the work surface as well as the serving area—you can pull up another chair for a third diner, in a pinch.

This project demonstrates plywood's ability to replace solid lumber. Note how the panel is laid out for cutting into sixteen leg slats—four per leg—plus a couple of drawer guides, ending up with about as close to zero waste as any cutting diagram can achieve. Note also, in the details, how these slats are then assembled with glue and finishing nails, pinwheel fashion, to create the legs; this makes for a lighter-weight table than if you used solid 4 × 4s for these legs—without any sacrifice in sturdiness.

Note also that since the drawer cutout will be used as the drawer front, it's probably quicker and easier to saw it out with cuts all the way across the panel. But unless you make these cuts with a very fine-toothed blade, you'll have to insert shim strips when you reassemble the four pieces, to compensate for the saw kerfs. The other option is to drill small holes at two opposite corners of the drawer front and cut it out carefully with a saber saw, using the finest blade you can, but leaving the framing panel intact (as shown on the panel layout). If you're going to surface the whole project with laminate, of course, any patch-up work will be concealed. In a photo, next page, strips of laminate are being applied to a leg assembly, masking the exposed plies.

You'll need scraps of ⅜" plywood for the drawer sides (which are rabbeted into the ends of the drawer front) plus another scrap

PANEL LAYOUT

¾" × **4'** × **8'**

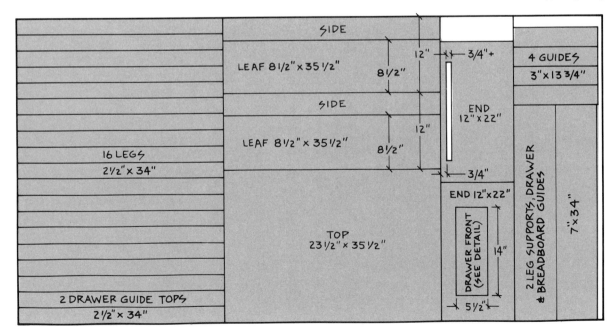

MATERIALS LIST

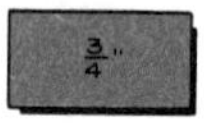
3/4"

Quantity	Description
1 4 x 8 panel	3/4" plywood, A-B or A-D Exterior
3 scraps	3/8" plywood, two 5 1/2" x 33 3/8" (for drawer sides) and one 5 1/2" x 12" (for drawer back)
4 pieces	1 x 1 lumber (3/4" x 3/4") x 34" for drawer and breadboard glides
1 piece	1 x 1 lumber x 10" for drawer handle (or use standard drawer handle)
2	#6 roundhead screws, 1 1/4" for drawer handle
2	Continuous hinges, 35 1/2" long with screws, for drop leaf installation
4	Folding brackets, 12" with screws, for leaf support
18 ft	Wood-pattern laminate (such as Conoflex) 30" wide, for table top, sides and legs
1/2 pint (approx)	Contact cement for installing laminate
1	Breadboard, 15 3/8" wide (cut from standard 18" x 24" breadboard)
As required	2d common brads for glue-nail construction of drawer; 4d casing or finishing nails for glue-nail construction of table; waterproof glue for glue-nail construction; wood dough for filling plywood edge voids and countersunk nail holes; fine sandpaper for smoothing filling and cut edges

DRAWER & BREADBOARD GLIDE ASSEMBLY

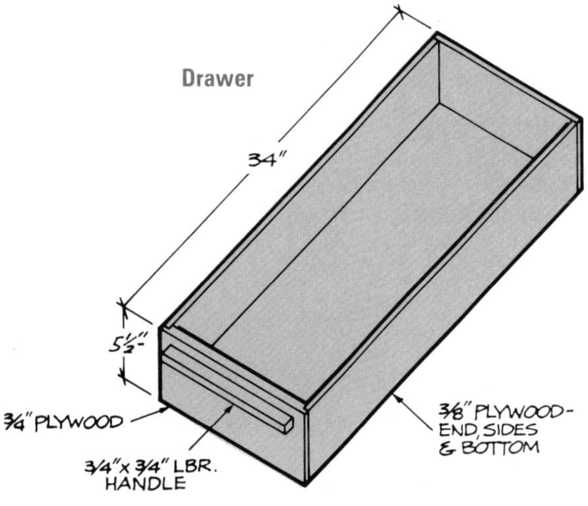
Drawer
34"
5 1/2"
3/4" PLYWOOD
3/4"x 3/4" LBR.
HANDLE
3/8" PLYWOOD-
END, SIDES
& BOTTOM

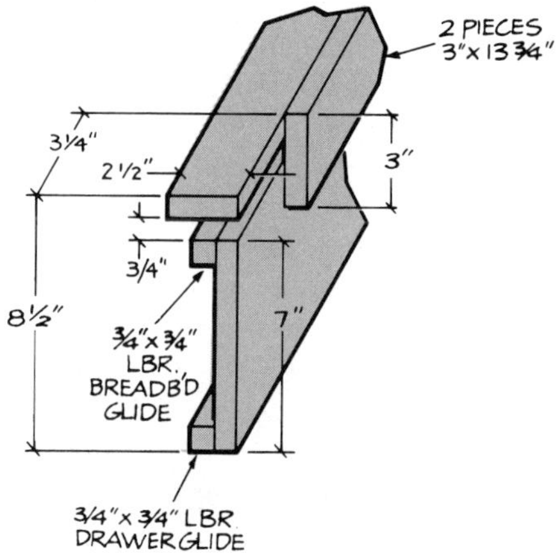
2 PIECES
3"x 13 3/4"
3 1/4"
2 1/2"
3"
3/4"
8 1/2"
7"
3/4"x 3/4"
LBR.
BREADB'D
GLIDE
3/4"x 3/4" LBR
DRAWER GLIDE

EXPLODED VIEW & DETAILS

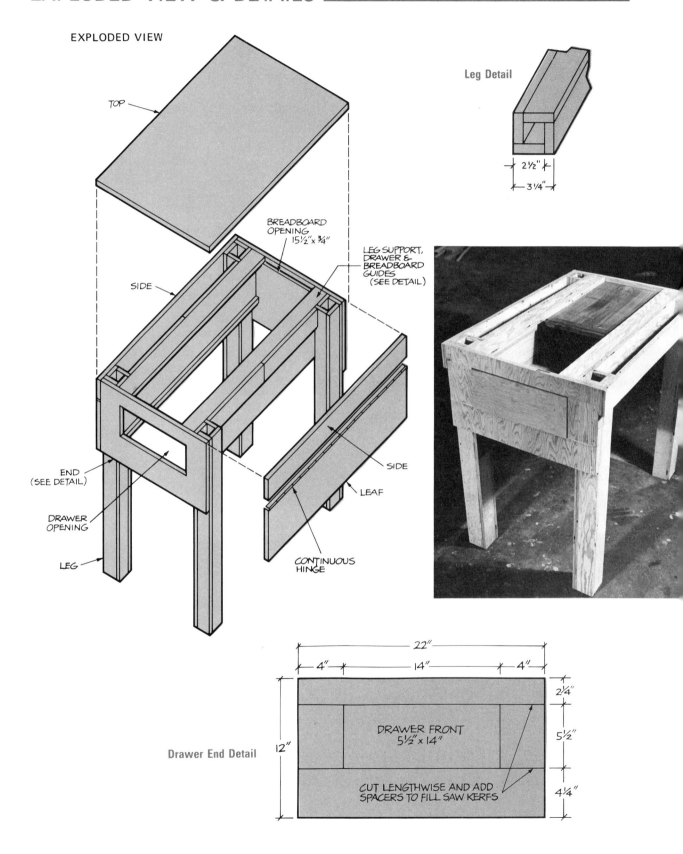

EXPLODED VIEW

Leg Detail

2½"

3¼"

TOP

BREADBOARD
OPENING
15½" x ¾"

LEG SUPPORT,
DRAWER &
BREADBOARD
GUIDES
(SEE DETAIL)

SIDE

END
(SEE DETAIL)

DRAWER
OPENING

LEG

SIDE

LEAF

CONTINUOUS
HINGE

Drawer End Detail

22"

4"   14"   4"

2¼"

DRAWER FRONT
5½" x 14"

5½"

12"

CUT LENGTHWISE AND ADD
SPACERS TO FILL SAW KERFS

4¼"

SHELF DETAIL

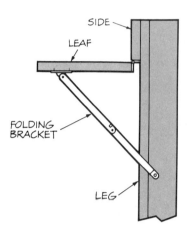

SIDE

LEAF

FOLDING
BRACKET

LEG

for the back and bottom. Though not shown this way in the detail sketch, I'd prefer to make the drawer bottom of a full-width piece of ¼" hardboard—the type with both faces smooth—overlapping the edges of both side and back pieces.

You'll also need ¾"-square wood strips for the drawer handle and glides, and for the breadboard glide. Wax these before installing. Finally, install the drop leaves on each side with a continuous hinge attached to the apron, and add folding support brackets at each leg position to hold these leaves level.

KNOCKDOWN PICNIC TABLE

A bench-table that comes apart and stacks nearly flat lets you toss it in your station wagon when you set off on picnics. Or, for patio dining, you needn't leave it set up for weather damage and soilage. With built-in seats on both sides, there's plenty of room for four adults. And it comes neatly out of a single panel of ¾″ plywood, with virtually no waste. (Note that no provision has been made for saw kerfs, so center your blade on the layout lines.)

Make your first cut on the 42″ mark to split the panel into manageable halves. Rip off the 6″ crosspieces and fasten them to the underface of the top and seats with steel angles, as shown. You can "nest" these crosspieces as shown in our second photo, for short-stack portage or storage. For extra strength, you'll also

David D. Gay
Kalamazoo, MI

probably want to attach the table supports to the sides permanently with glue and screws. The center crosspiece for the table has double-width slots to fit down over this assembly. You've now created five units that quickly assemble and come apart,

MATERIALS LIST

Quantity	Description
1 4 x 8 panel	¾" plywood, MDO or A-C Exterior
6	3" x 5" angle braces with screws
6	3" x 3" angle braces with screws
As required	3d finishing nails, resorcinol type (waterproof) glue, wood dough or synthetic filler, fine sandpaper, top-quality acrylic latex paint and stain resistant companion primer

PANEL LAYOUT

¾" × **4'** × **8'**

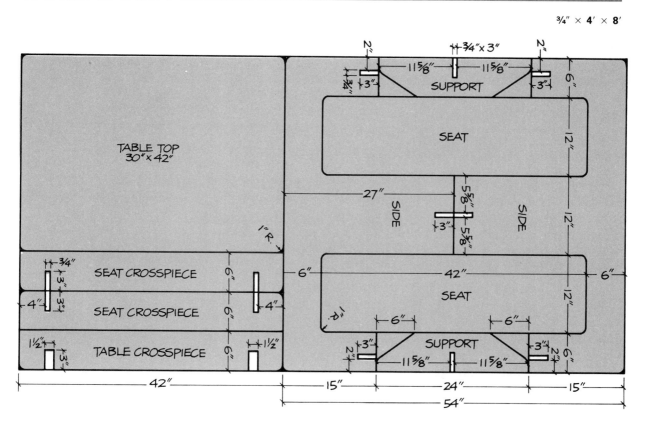

yet—if you do a precision job with your slots—will be unusually sturdy when it's set up.

About those slots: Allow for a bit more thickness so that when you apply your outdoor primer and topcoat the mating pieces won't bind.

We finished our table in a light milk-chocolate brown, but enameled all edges white. If you plan to accent your edges in a similar way, be certain to inspect them carefully before finishing. Fill all voids with wood dough (or small glue blocks) and sand them smooth before priming.

FRONT VIEW

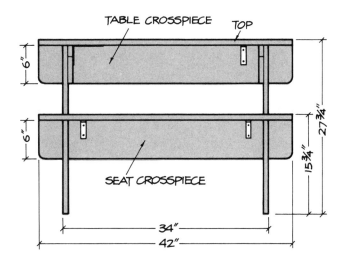

SIDE VIEW

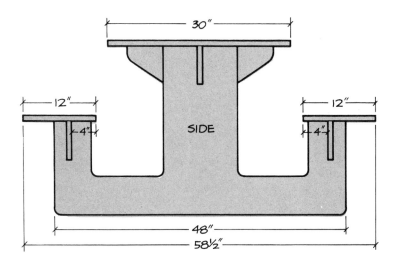

EXPLODED VIEW: SEAT ASSEMBLY

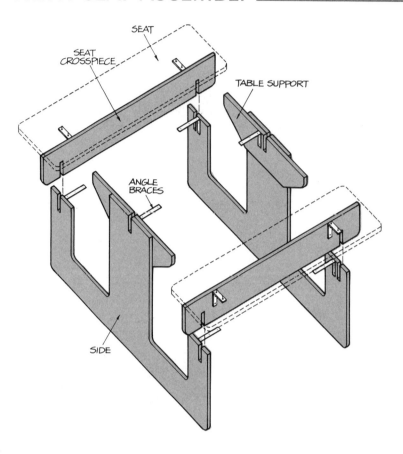

SEAT

SEAT
CROSSPIECE

TABLE SUPPORT

ANGLE
BRACES

SIDE

EXPLODED VIEW: TABLETOP ASSEMBLY

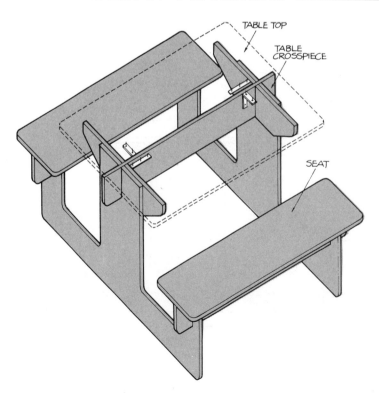

TABLE TOP

TABLE
CROSSPIECE

SEAT

FOLD-FLAT PLAY TABLE

Little wonder this fold-up-and-tote wonder won a recent Grand Prize in the POPULAR SCIENCE competition. It has the intriguing look of a 3-D puzzle, and it's just about as much fun to put together. Truly, here's a project you'll enjoy making. And it all comes out of a half-sheet of ¾″ plywood.

We turned our prototype into a three-tone sculpture by enameling one side of the panel a creamy yellow, the other side purple, and all the edges white. The finishing job was made a lot simpler by using MDO plywood. One overall prime coat was all the surface preparation that was needed (aside from sanding all cut edges).

Note that the cuts are numbered on the layout diagram (also note the best direction for the face grain). Start with the two seat

Charles S. Kocher
Newton, IL

cutouts, drilling holes in the two upper corner areas—just large enough to insert your fine-toothed saber-saw blade. Make the short "No. 1" cut to the square corner; stop and back out the blade. Re-insert in the same hole and make the other "No. 1" cut, then move the blade to the second hole and make Cut 2 to meet the first cut at the square corner. Re-insert the blade in the second hole and complete the seat cutout by making Cut 3. Repeat for the second seat, then split the panel in two parts by making Cut 4 with either your saber saw, run along a clamped straight-edge, or—for faster cutting—a portable circular saw.

Drill holes as shown in the slot waste and cut out the table with the sequence of cuts indicated. When all separations are made, sand all edges, rounding corners slightly, do all painting—and now the fun begins! Reassemble all parts like a giant jigsaw puzzle and apply half of the butt hinges to the top surface, as indicated. Flip the reassembled panel over and apply the other eight hinges. Now apply the six barrel bolt latches, and the turn-buckle-adjusted hook for securing the set-up unit.

PANEL LAYOUT

3/4" × 4' × 4'

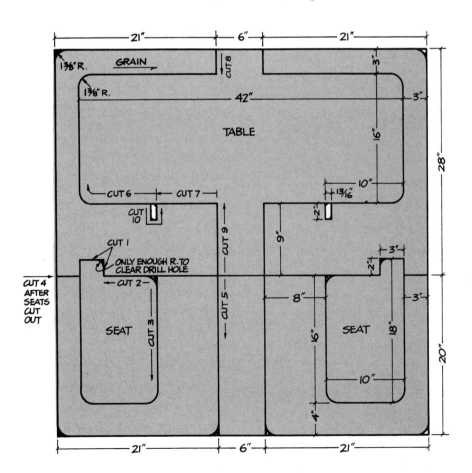

MATERIALS LIST

Quantity	Description
1 4 x 4 panel	¾" plywood, MDO or A-B or A-C Exterior
16	2½" x 1¾" butt hinges w/screws
6	3" barrel bolt latches w/screws
1	Turnbuckle (11" reach) w/3 screw eyes
2 each	¼" x 1½" bolts, ¼" x ⅞" nuts, ¼" wing nuts (wing nut adjustments reinforce turnbuckle tension and permit table leveling)
13	Furniture bumpers
	Velcro (approx. 1" x 3") w/wood strip spacer (approx. ¼" x 1" x 3")
As required	Wood dough or synthetic filler, epoxy, fine sandpaper, top quality finish

TOP VIEW

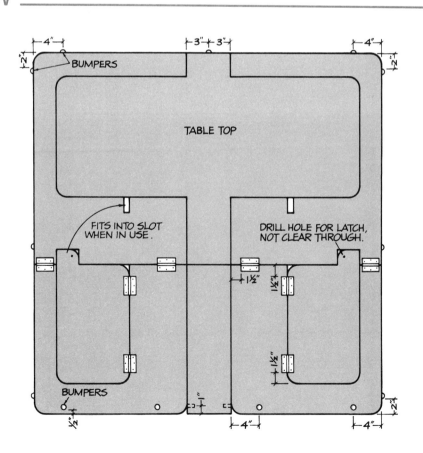

BUMPERS

TABLE TOP

FITS INTO SLOT WHEN IN USE.

DRILL HOLE FOR LATCH, NOT CLEAR THROUGH.

BUMPERS

UNDERSIDE

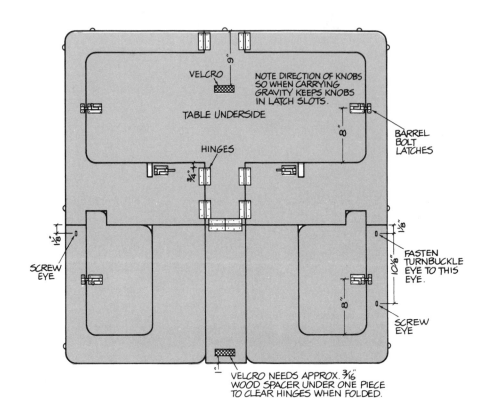

VELCRO

NOTE DIRECTION OF KNOBS
SO WHEN CARRYING
GRAVITY KEEPS KNOBS
IN LATCH SLOTS.

TABLE UNDERSIDE

BARREL
BOLT
LATCHES

HINGES

SCREW
EYE

FASTEN
TURNBUCKLE
EYE TO THIS
EYE.

SCREW
EYE

VELCRO NEEDS APPROX. 3⁄16"
WOOD SPACER UNDER ONE PIECE
TO CLEAR HINGES WHEN FOLDED.

BACK VIEW

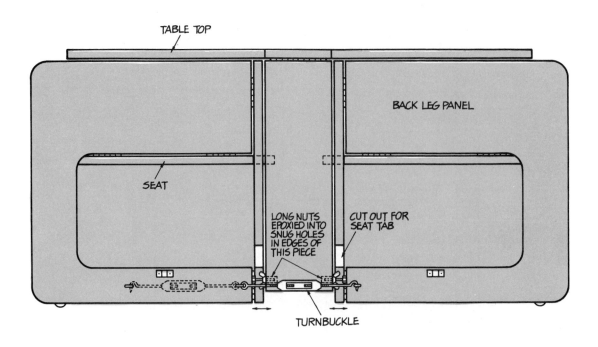

TABLE TOP

BACK LEG PANEL

SEAT

LONG NUTS
EPOXIED INTO
SNUG HOLES
IN EDGES OF
THIS PIECE

CUT OUT FOR
SEAT TAB

TURNBUCKLE

SOLAR DEHYDRATOR

Once you've loaded up the removable expanded-metal shelves, you clamp on the collector-cover, carry the unit outdoors, pivot the cover down for best solar exposure—and thermal-convection air flow dries the food for storage without refrigeration.

This cleverly designed dehydrator can produce a pantryful of tasty dried fruits and vegetables. And it takes just one half-sheet of ¾″ plywood and a few other inexpensive materials to build it. Since it's "powered" by solar energy, it costs nothing to use. It depends on the movement of air, heated when the sun strikes the black-painted collector plate, thermosyphoning up through the food-laden wire shelves. The inlet and outlet openings for this air flow are screened to keep out insects during the drying

Dave Fullmer
Salt Lake City, UT

process. Note that the collector's screening can be held in place with a rubber strip pressed in a groove cut around the edge.

Basically the unit consists of two boxes with acrylic "lids." The top and bottom of the shelf box are not permanently fastened, but are cleverly notched to stay in place (the bottom slides in a groove, the top friction-fits over the projecting handlegrips). The pivot is a simple disk rotation. Note that the woodscrews—four to a side—are not fully driven, so their heads protrude enough for the luggage catch to hook over. The top screw on each side is the latch for the vertical (closed) position of the collector cover. The other give three angles of adjustment for best solar exposure.

Paint all plywood parts gloss black and operate your unit without food for a few days to allow full curing of the paint. When you load the unit (single layers on each shelf) and place it in the sun, check every two hours, at first, and turn the slices to facilitate uniform drying. When the sun is gone, cover the screened vents with plastic to keep heat in and carry the unit indoors. The drying process can be continued next morning. Most foods will take two to three sunny days to dry enough for storage.

PANEL LAYOUT

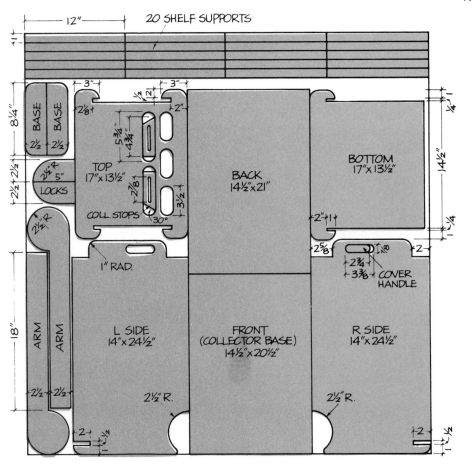

EXPLODED VIEW

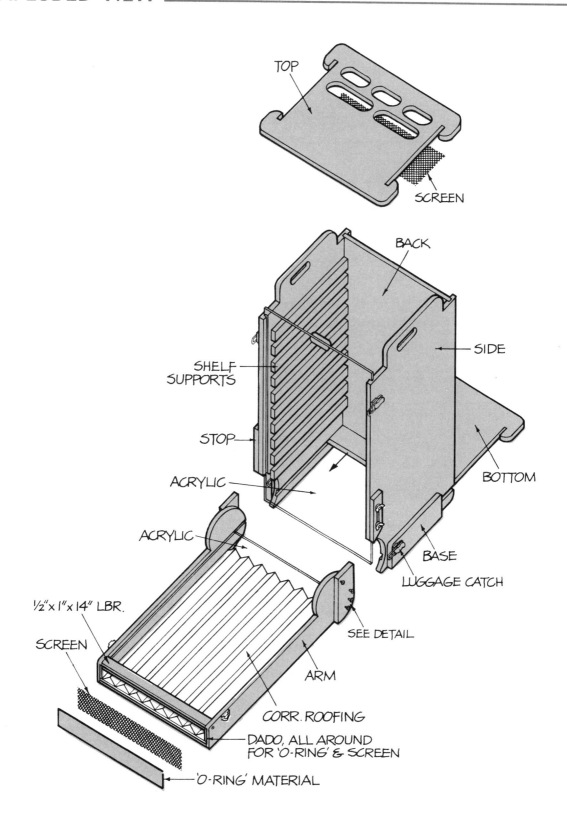

TOP

SCREEN

BACK

SIDE

SHELF
SUPPORTS

STOP

ACRYLIC

ACRYLIC

BOTTOM

BASE

LUGGAGE CATCH

SEE DETAIL

½" x 1" x 14" LBR.

SCREEN

ARM

CORR. ROOFING

DADO, ALL AROUND
FOR 'O-RING' & SCREEN

'O-RING' MATERIAL

MATERIALS LIST

Quantity	Description
1 4 x 4 panel	¾″ plywood, MDO
5 sq. ft.	⅛″ acrylic sheet (or tempered glass)
10 pcs.	13⅞″ x 12″ anodized expanded metal, or as desired (for shelves)
1	12″ x 16″ fine mesh aluminum window screen (cut to appropriate sizes)
4	Luggage-type catches
4	Screws and wing nuts
1	13″ x 18″ aluminum or galvanized corrugated roofing material
1	14″-long foam seal
Approx. 36 lin. in.	"O-ring" material
As required	Wood screws, white or urea resin glue, flat black exterior enamel paint

ARM HINGE DETAILS

Detail A (see Exploded View)

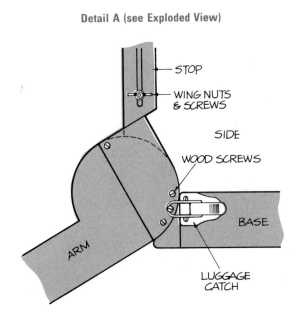

STOP
WING NUTS & SCREWS
SIDE
WOOD SCREWS
BASE
ARM
LUGGAGE CATCH

Interior of Detail A

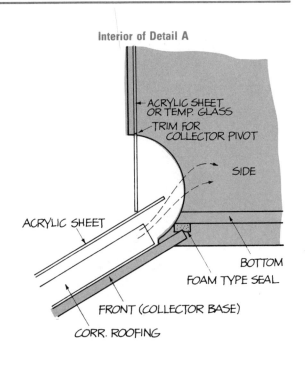

ACRYLIC SHEET OR TEMP. GLASS
TRIM FOR COLLECTOR PIVOT
SIDE
ACRYLIC SHEET
BOTTOM
FOAM TYPE SEAL
FRONT (COLLECTOR BASE)
CORR. ROOFING

DETAILS

Dado details for bottom and back

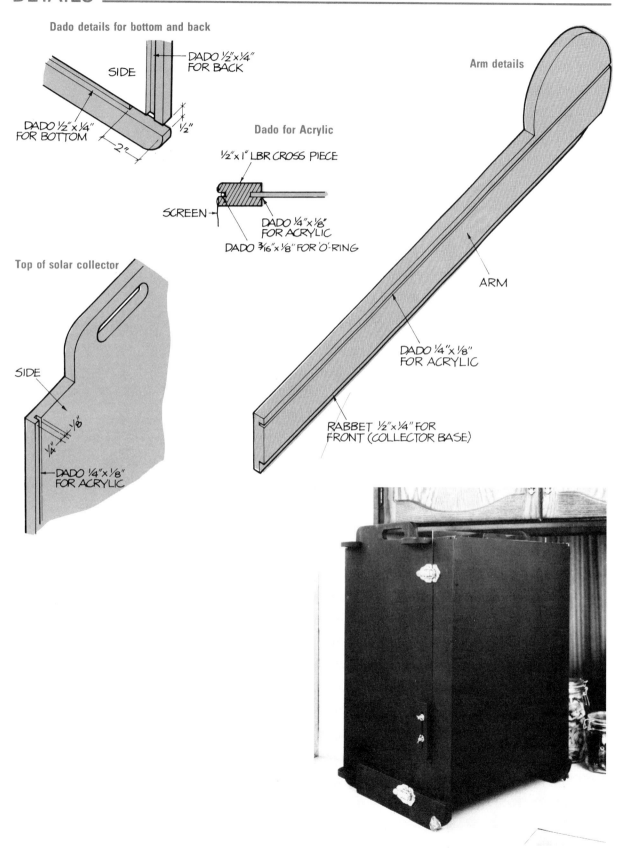

SIDE

DADO ½″ x ¼″
FOR BACK

DADO ½″ x ¼″
FOR BOTTOM

½″

2″

Arm details

Dado for Acrylic

½″ x 1″ LBR CROSS PIECE

SCREEN

DADO ¼″ x ⅛″
FOR ACRYLIC

DADO ³⁄₁₆″ x ⅛″ FOR 'O'-RING

ARM

DADO ¼″ x ⅛″
FOR ACRYLIC

Top of solar collector

SIDE

⅛″

¼″

DADO ¼″ x ⅛″
FOR ACRYLIC

RABBET ½″ x ¼″ FOR
FRONT (COLLECTOR BASE)

BARBECUE SIDE CART

Want to know why this has always been one of my favorite projects—even though it dates from the very first year of our contest? Study that panel layout on the next page: The *only* waste is the sawdust from the kerfs! The circular cutouts are put to use as wheels, feet and shelves. Note that the two side panels are positioned front-to-front at the lower right of the layout, so you can scribe their "shared circles" more easily. (The dotted lines on the layout indicate shelf positions.)

Don't suppose that, since the unit is sleekly narrow, it's a pushover. That bottom tray is intended for heavy items, like sacks of briquets, which provide ballast. The upper shelves are an ideal caddy for all the gear a barbecue chef needs—but this unit isn't

John Allan Jones
Santa Rosa, CA

limited to that function. It can also serve as a wheel-out bar for the deck or patio, with liquor and mixers stored on the tray below, glasses, ice bucket and finger food on the upper shelves.

Although it's possible to assemble this whole unit (with certain dimension adjustments) using simple butt joints and screws, since it's only ⅝″ plywood you'll get a stronger assembly if you rout ⁵⁄₁₆″-deep grooves for the shelves, as shown, then assemble with glue and finishing nails. Note the bottom tray assembly is beefed up for its heavy duty with steel corner braces.

If you go the butt-joint route, use 1 × 1 lumber strips or milled moldings as ledgers beneath all shelves, as shown in the details, next page.

To avoid complexity, the cutting layout makes no provision for saw-kerf waste, so center all separation cuts on the layout lines.

You'll need to buy a 20½″ steel rod, ¼″ in diameter and with holes drilled through both ends, to serve as the axle. Cable clamps make good straps to fasten the rod on the underside of the cart, but you'll probably have to make the forked wheel stops from metal angles. With a washer on both sides of the wheel and cotter pins through both holes, you should have a cart that's fairly easy to roll the short distances you'll be moving it.

Those half-disk feet can simply be glued and nailed to each side of the tray, but in our exploded view you'll see that we rabbeted them across the top for greater strength.

PANEL LAYOUT

⅝″ × 4′ × 8′

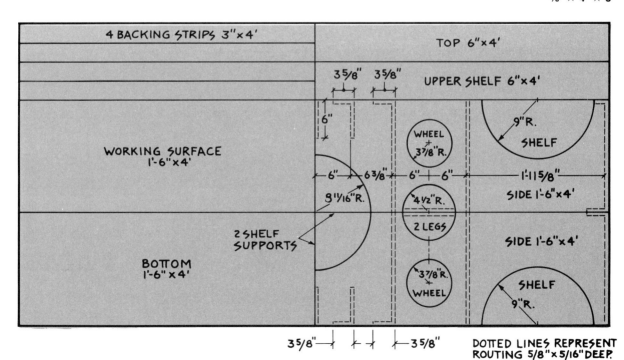

■ MATERIALS LIST ■

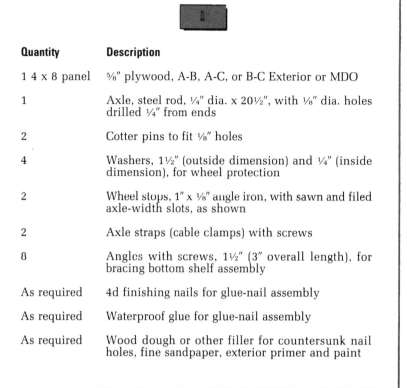

Quantity	Description
1 4 x 8 panel	⅝″ plywood, A-B, A-C, or B-C Exterior or MDO
1	Axle, steel rod, ¼″ dia. x 20½″, with ⅛″ dia. holes drilled ¼″ from ends
2	Cotter pins to fit ⅛″ holes
4	Washers, 1½″ (outside dimension) and ¼″ (inside dimension), for wheel protection
2	Wheel stops, 1″ x ⅛″ angle iron, with sawn and filed axle-width slots, as shown
2	Axle straps (cable clamps) with screws
8	Angles with screws, 1½″ (3″ overall length), for bracing bottom shelf assembly
As required	4d finishing nails for glue-nail assembly
As required	Waterproof glue for glue-nail assembly
As required	Wood dough or other filler for countersunk nail holes, fine sandpaper, exterior primer and paint

SHELF DETAILS

Shelf Support Dimensions: Detail A

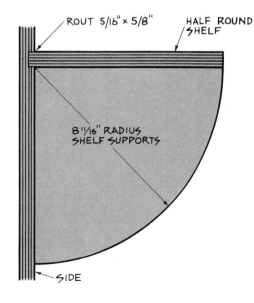

Alternate Assemblies, Upper Shelf: Detail B

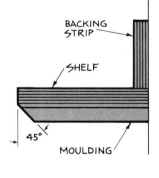

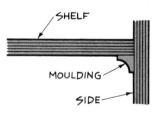

SHELVES SUPPORTED BY MOULDINGS INSTEAD OF ROUTING

EXPLODED VIEW & WHEEL DETAIL

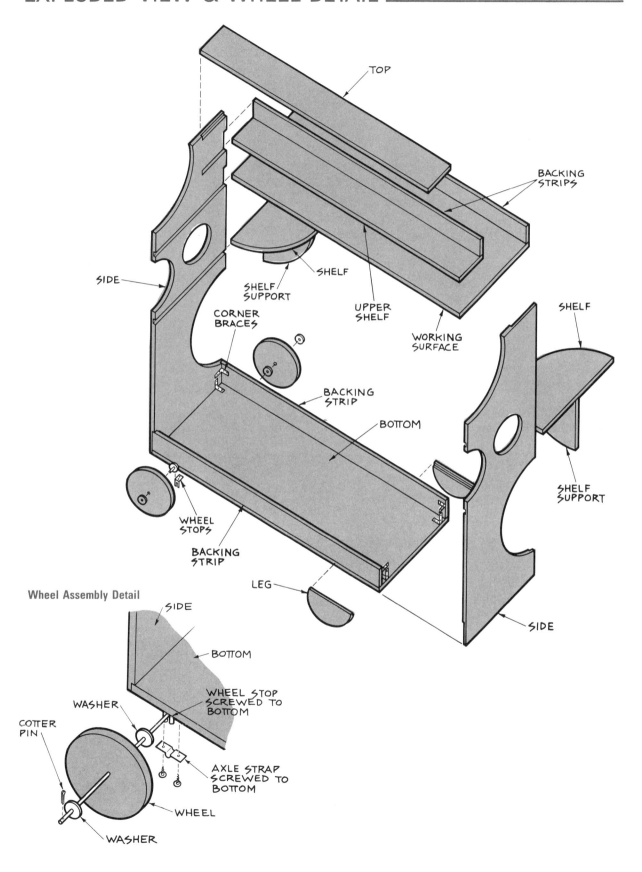

TOP

BACKING
STRIPS

SIDE

SHELF

SHELF
SUPPORT

CORNER
BRACES

UPPER
SHELF

WORKING
SURFACE

SHELF

BACKING
STRIP

BOTTOM

SHELF
SUPPORT

WHEEL
STOPS

BACKING
STRIP

LEG

SIDE

SIDE

Wheel Assembly Detail

SIDE

BOTTOM

WHEEL STOP
SCREWED TO
BOTTOM

WASHER

COTTER
PIN

AXLE STRAP
SCREWED TO
BOTTOM

WHEEL

WASHER

BARBECUE COUNTER-CART

Mounted on casters, this barbecue counter can be positioned anywhere on deck or patio to serve as work table, bar or buffet. The ⅝″ panel is Texture 1-11 exterior siding, cut and applied to a 2×4 frame. The base, center partition and counter top can be cut from a standard ¾″ panel of A-C exterior plywood—or you can cut these from the same siding, using it (in the case of the countertop) as a base for applying plastic laminate. If you choose to use the ¾″ ply, you'll have a thickness discrepancy in piecing out the back panel since, as you'll see from the panel layouts, the front and back are each in two pieces. These pieces are positioned on the panels to take advantage of the rabbeted long edges of T 1-11 panels: When you butt opposite edges together, they

Richard W. Riker
Grants Pass, OR

automatically form a groove that matches the spaced striations that give the panel its name.

Note that most of the two end panels are hinged to provide access to the storage compartments. The other panels are simply nailed to a simple butt-joint frame of 2 × 4s. The two pieces that form the off-center divider must have their upper corners notched to take the horizontal 2 × 4s. You can position this divider to accommodate specific storage. In our prototype, the deeper com-

PANEL LAYOUTS

19/32″ × **4**′ × **8**′

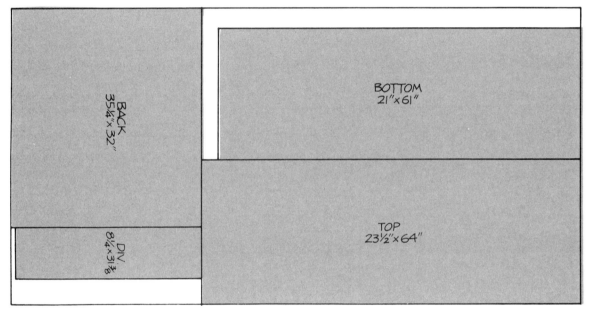

19/32″ × **4**′ × **8**′

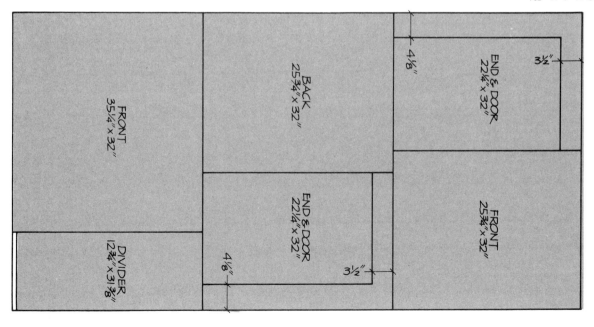

MATERIALS LIST

Quantity	Description
2 4 x 8 panels	$^{19}/_{32}$" Texture 1-11 siding, or other plywood siding
30 lin. ft	2 x 4 framing, cut into following lengths: 58" (2), 27¾" (6), 21" (2), 18" (1)
1 sheet	24½" x 65" formica (for top of stand)
2 each	Door handles, magnetic door catches
2 pair	Butt hinges
4	3" ball-type casters
4 pieces	½" x 1¼" for top edge trim; 2 cut 24½" long, 2 cut 65" long
As required	10d common nails (for assembling framing); 8d nonstaining box, siding, or casing nails (for fastening plywood to framing); top-quality oil-base semi-transparent stain, oil or latex solid color stain, or acrylic latex paint with stain-resistant companion primer

Note: Texture 1-11 and other such sidings are manufactured with
   shiplapped edges so that grooves occur at edge joints. Be
   sure to follow panel layout and assembly drawings
   carefully to assure proper edge joints.

FRONT & END VIEW

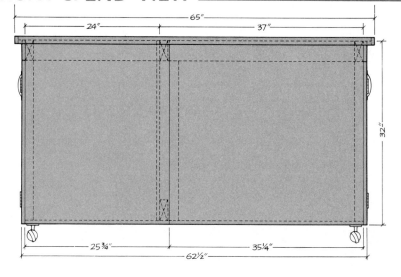

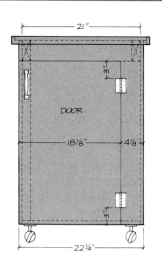

partment lets us stow folding aluminum chairs; the shallower end takes bags of charcoal and cooking utensils.

The assembled cabinet measures about 5' long by 2' wide and 3' high, so be sure you have ample space to roll it into when the party's over.

As for a finish, you'll probably just want to apply a semitransparent strain to the siding—especially if it's a rough-sawn texture. But you'll need a more durable surface for the countertop. In our case, we applied plastic laminate and cemented strips of it to edges we'd beefed up with mitered ½"-thick trim.

EXPLODED VIEW

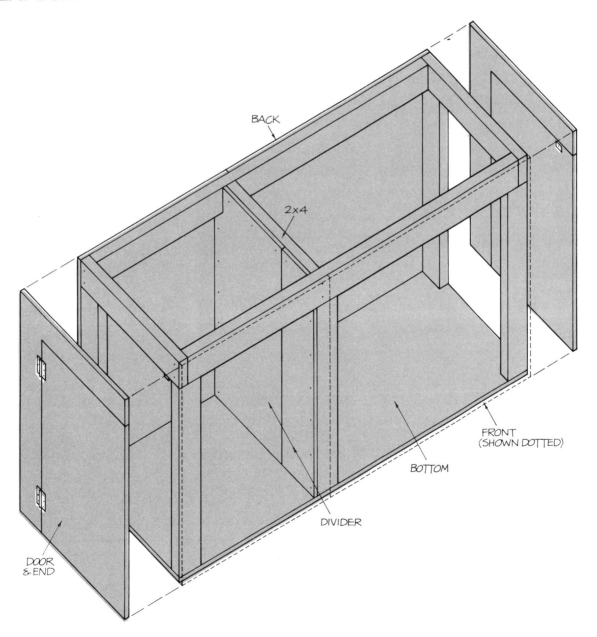

HOUSE-NUMBER LAMP POST

Identify your home elegantly while lighting a welcome for visitors with this unique personalized lamp post. It welcomes feathered visitors, as well, with a built-in birdhouse complete with perch—and a boxed shelf sports a flowering potted plant. These units share a platform that's pierced by the 4 × 4 post that holds the entire assembly off the ground. The little name plate, swinging from a 1″ dowel, is an optional final touch.

After cutting all parts (centering your kerfs on the layout lines), miter all vertical-joint edges 45 degrees—especially important if you plan to put a natural weatherproof finish on the assembly, as in the photograph: You don't want exposed plies at the corners. Attach all inside pieces (*H, J, Q* and one of the triangular *F* brack-

Don Sepulski
Sarasota, FL

ets) to the 4×4 post first, then assemble the box around these parts. Install the back (U) and its stiffeners, then sides (N, O) and pieces A and I. Run the electrical cable through this assembly, fastening it with cable clamps as shown. Attach it to an outlet box and light socket. Add the three plastic windows (the prototype features smokey or bronze Plexiglas) and the remaining top pieces (G, R, S and T), gluing all joints with waterproof glue and securing with finishing nails.

Lay out your house number on panel K and cut it out with a scroll saw. Install the panel of translucent white plastic on back; it will "borrow" light from the windowed light above. Note it is "cleated" in place with parts D and E, but be sure to drill clearance holes through it before driving the screws.

Add front panels K and M to the assembly. Glue-nail top piece G in place after coating edges with caulk to seal moisture out of the lamp area. The light bulb can be changed later by removing one of the window pieces, or (if you prefer) by making a hinged door in back panel U.

If you have not used a pressure-treated 4×4 (or redwood or cedar) for the post, soak the portion that will be underground in a liquid preservative. Depending on soil conditions, you'll want to bury up to four feet, backfilling firmly. If you plan to anchor the post in concrete, two feet is adequate, shortening the total length accordingly. The electric cable should be of a buryable outdoor type; run it down the back side of the post and through a trench to your home.

PANEL LAYOUT

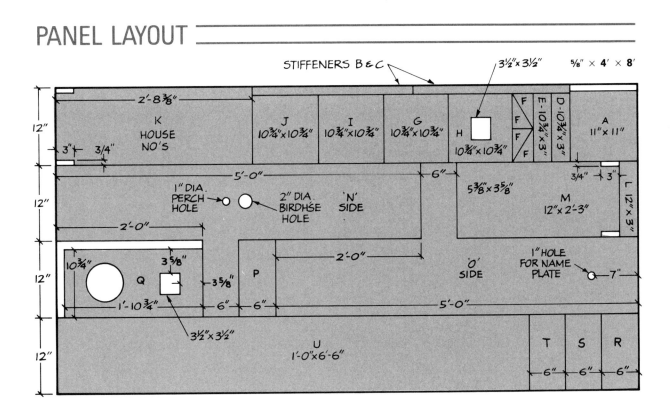

MATERIALS LIST

Quantity	Description
1 4 x 8 panel	⅝" plywood, A-C Exterior, 303® Exterior siding, or MDO
1	Post, 4 x 4, 8' long for standard
1	Dowel, 1" dia. x 2' long for bird perch and name plaque
1 piece	Plexiglas (or similar material), translucent white, 10" x 20" x ⅛" thick for lighted house number
1 piece	Plexiglas (or similar material), clear or bonze transparent, 11" x 14" x ⅛" thick for lantern front
2 pieces	Plexiglas (or similar material), clear or bonze transparent, 14" x 11⅛" x ⅛" thick for lantern sides
1 each	4" outlet box, ceramic or porcelain light socket (with mounting screws) plus large light globe
1	Outdoor underground electrical wire, #12 or #14, depending on local code. Length as required.
3	Cable clamps to attach light wire to inside back of box
As required	Wood screws, flathead, #8, 1" long for glue-screw attachment of stiffeners B and C, and backing strips D and E
As required	Wood screws, roundhead, #4 or #6, ½" long, for lantern and number plate Plexiglas attachment
1	Name plaque with hooks and eyes for attachment
1	Flower pot, up to 8" dia. for planter
As required	Waterproof glue for glue-nailing joints
As required	Nails, 3d, galvanized casing or finishing for glue-nailing
As required	Wood dough or other filler for countersunk nail holes, if desired; caulking for sealing around Plexiglas edges; fine sandpaper for smoothing cut edges; paint or stain for finishing.

SIDE "N" CUTAWAY

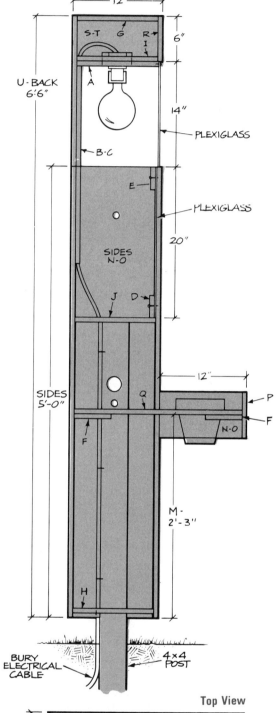

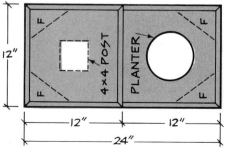

FRONT VIEW

ISOMETRIC VIEW

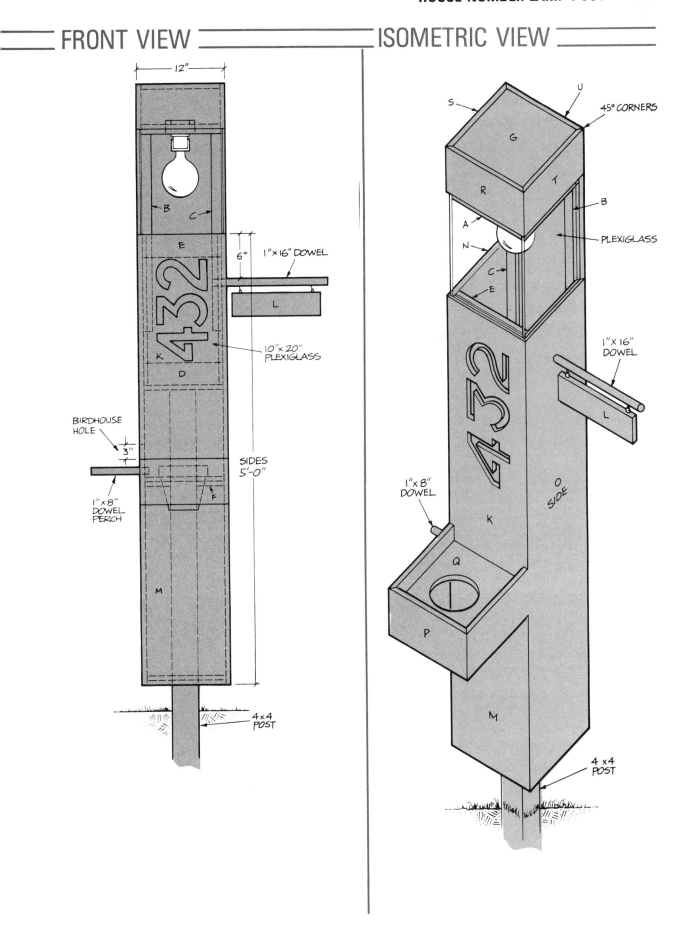

FRONT VIEW labels:
12"
B
C
E
6"   1"x16" DOWEL
432
L
K   10"x20" PLEXIGLASS
D
BIRDHOUSE HOLE
3"
SIDES 5'-0"
1"x8" DOWEL PERCH
F
M
4x4 POST

ISOMETRIC VIEW labels:
S
U
45° CORNERS
G
R   T
A   B
N   PLEXIGLASS
C
E
432
1"x16" DOWEL
L
1"x8" DOWEL
O SIDE
K
Q
P
M
4x4 POST

ARMILLARY SPHERE SUNDIAL

Creating a round project from a flat, square sheet inevitably means some waste, but the nesting of parts on this half-sheet of ¾" MDO is still shrewdly frugal. They assemble into a functional sundial with a built-up hour ring; none of the parts are bent.

Begin your layout by marking diagonal lines, corner to corner, as indicated in dotted lines on the panel layout, next page. These lines serve as compass points for all circles and segments. On one side of rings G and H, make small ⅛" notches at inner edges along diagonal lines, to be used as drilling guides. Also notch the centerline of spacers J, and the top of the center of the vertical base side supports (B).

You can use a length of flexible wire with a nail fastened to

Sharlene Landers
Norwalk, CT

one end and a pencil to the other to scribe all circles and arcs. Draw all the same-size curves at one time, starting in the center with the largest 36″-dia. circle (an 18″ length of wire). Shorten the wire accordingly for each smaller circle. Some compass points are used more than once. Now cut out all pieces with a saber saw—or carry the panel to a full-size bandsaw.

Start assembly with the base. Lay the top and bottom interior supports (D and E) on one of the segments B as shown in a detail. Glue-nail in place using 3d finishing nails, then trim off crescent tips of D as shown. Glue-nail the other B segment in place with 6d finishing nails. Position this assembly in the center of main base A and screw together from underneath. Position base supports C at right angles to this assembly and screw from underneath; also toe-nail the C pieces to the assembly. If the C pieces extend beyond the base, cut them flush with the edge. After

PANEL LAYOUT

¾″ × **4′** × **4′**

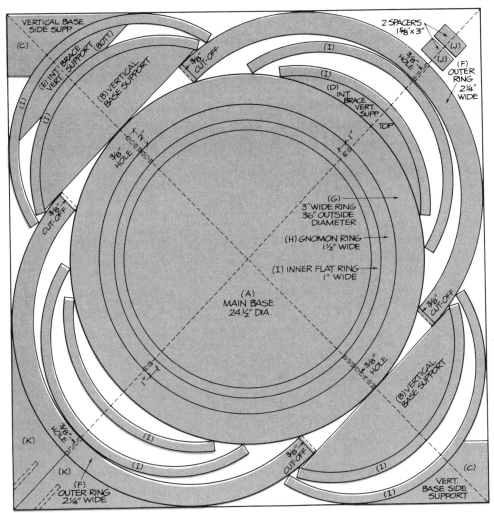

MATERIALS LIST

Quantity	Description
1 4 x 4 panel	¾" plywood, MDO or, if available, A-B Exterior
49 in.	⅜" dia. aluminum rod (for gnomon rod and to join rings F and H)
1 set	¾" self-adhesive vinyl letters (for compass face)
1 set	2" self-adhesive vinyl numbers (to indicate hours on ring I and degree of latitude on ring G)
As required	3d and 6d finishing nails, 1¼" and 1¾" #8 flathead wood screws, resorcinol type (waterproof) glue, wood dough or synthetic filler, fine sandpaper, top-quality acrylic latex exterior paint and stain resistant companion primer

F & G RINGS: ASSEMBLY

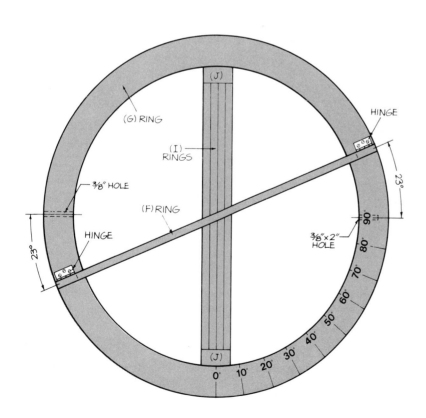

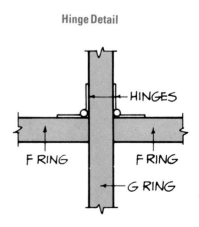

Hinge Detail

painting this entire base, you can design a compass face using self-adhesive vinyl letters as shown in the top view.

The hour ring is assembled from four layers of *I* pieces glue-nailed into a stack. Start with inner flat ring *I*, a full circle. Add two long *I* sections and part of one short *I* section, trimmed to fit. For layer 3, use three short *I* sections plus part of the above scrap. Layer 4 is two long sections plus another scrap.

Using the notches as guides, drill two ⅜"-dia. holes 180 degrees opposite one another in ring *G*. Drill one hole through, but the other only 2" deep. Glue and toenail spacers *J* to ring *G* on centerline notches at right angles to the holes. Using the spacer centerline as zero and the blind hole as 90 degrees, divide each quarter circle into nine 10-degree segments as sketched.

Set ring *G* in the base and place ring *I* on the spacers. Mark the centerpoint where ring *I* meets the spacers; remove the ring and drill pilot holes for 1¾" #8 flathead wood screws. Using one screw hole to indicate 12 noon, divide the ring into quarters (use the notches as guides).

For gnomon ring *H*, again using the notches as guides, drill two ⅜" holes opposite one another, through the ring. Now place ring *I* inside ring *H* and put them inside ring *G*. Run the 42"-long aluminum rod through the top of ring *G*, through both sides of

I RING

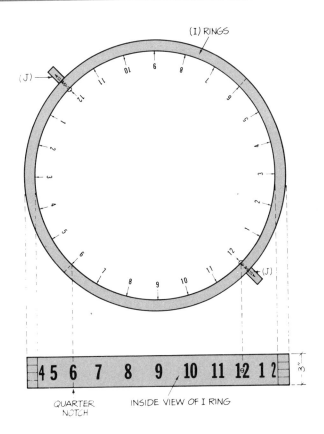

INSIDE VIEW OF I RING

QUARTER NOTCH

ring *H* and the center of ring *I* and into the blind hole of ring *G*. (To prevent wobble, you can put washers on the rod between rings *G* and *H*.)

Glue two *K* pieces together to make the rod ornament, drilling it as indicated. Set ring *H* at right angles to ring *G* and position ring *I* for gluing and screwing to the spacers (*J*). Cut ⅜″ off the ends of both halves of ring *F* to take the width of ring *G*. To find the points where ring *F* connects, measure up 23 degrees from the 90-degree mark on ring *G*, then go directly across from the 90-degree mark to the ⅜″ hole and measure down 23 degrees (see detail drawing). Attach a hinge to each end of both halves of ring *F* and position each half so that the center of the ring hits the 23-degree marks. Screw hinges to ring *G*. Make sure rings *F* and

ISOMETRIC VIEW

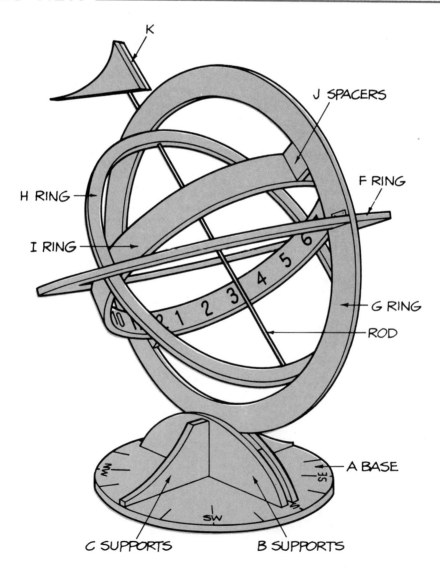

BASE DETAILS

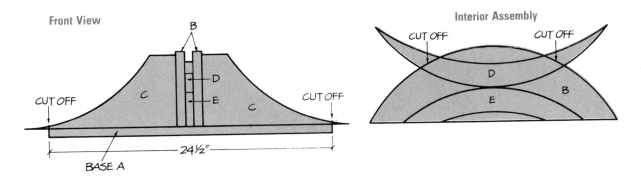

Front View

B

CUT OFF

D

E

C   C

CUT OFF

CUT OFF

24½″

BASE A

Interior Assembly

CUT OFF   CUT OFF

D

E   B

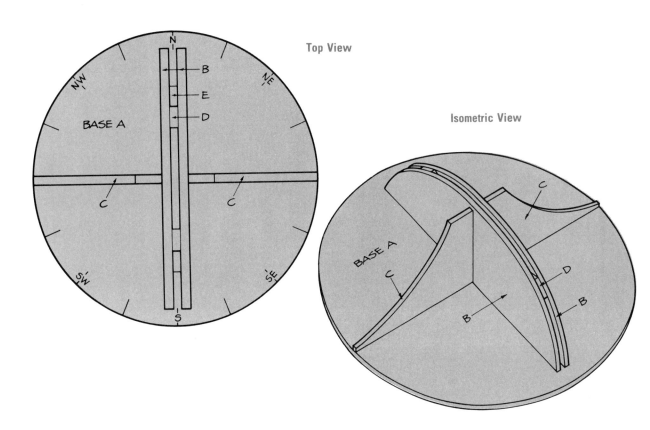

Top View

N

B

NW   NE

E

D

BASE A

C   C

SW   SE

S

Isometric View

BASE A

C

C

D

B   B

H are at right angles, then drill a ⅜″-dia. hole through both sides of ring *F* and ¾ of the way through both sides of ring *H*. Cut two pieces of the rod 3¼″ long and insert in these holes.

You'll have to orient your sundial to your latitude (which you can look up on a map or in an almanac). Slide the sundial in the base until your latitude degree lines up with the center notch. Then, at noon, put the sundial in the sunlight with the gnomon rod pointing north. Turn the sundial until a shadow from the rod falls on the 12-noon position of Ring *I*. When properly positioned, secure ring *G* to the base by driving a 1¼″ # 8 FH wood screw at either side of the *C* support.

LAWN SCULPTURE

The only piece of non-functional sculpture ever to win an award in our design competion, this handsome lawn piece was first created as a cardboard model for a senior art-class competition. When it won, the student undertook construction at home as a graduation gift to her high school before she left for college. It is assembled from four sheets of ½″ plywood—two of each layout shown—with the panels butt-joined via glue-blocks, as shown in the diagram. Note that large-numbered panels mate with smaller-numbered edges.

The two identical assemblies have their "H" sections locked together, their faces offset about 2″. This can be a friction fit or the joint can be glued. The prototype sculpture was painted bright

Susan Wagner
River Vale, NJ

yellow with a good grade of exterior enamel. But if you prefer, you could paint each a contrasting color before joining them, or even use contrasting two-tone effects, with surfaces one color and edges another. Be certain to use exterior glue and seal all edges before painting. For sleek painted surfaces like these, you're ahead choosing MDO plywood.

PANEL LAYOUTS

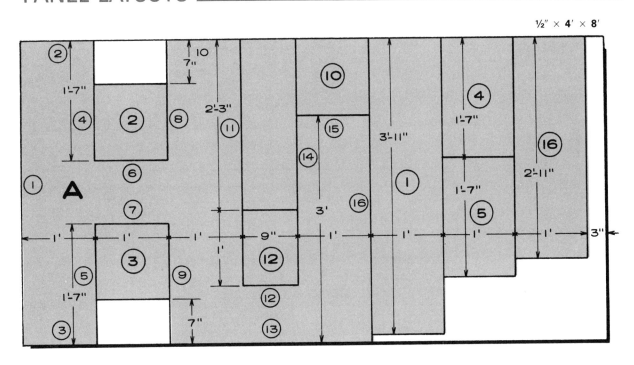

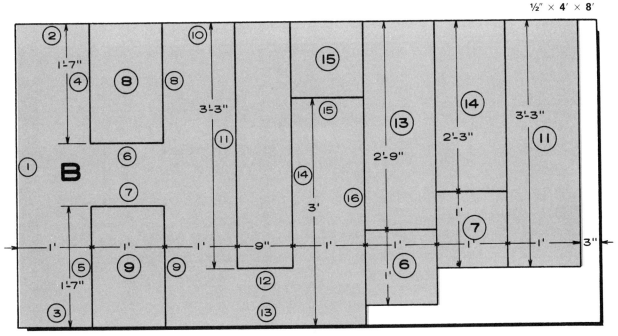

MATERIALS LIST

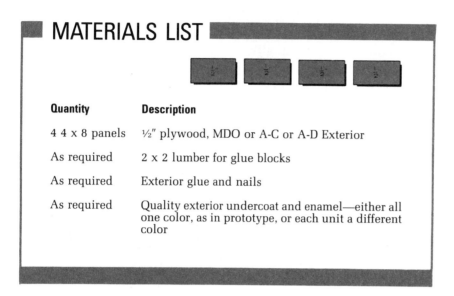

Quantity	Description
4 4 x 8 panels	½″ plywood, MDO or A-C or A-D Exterior
As required	2 x 2 lumber for glue blocks
As required	Exterior glue and nails
As required	Quality exterior undercoat and enamel—either all one color, as in prototype, or each unit a different color

BRACING DETAILS

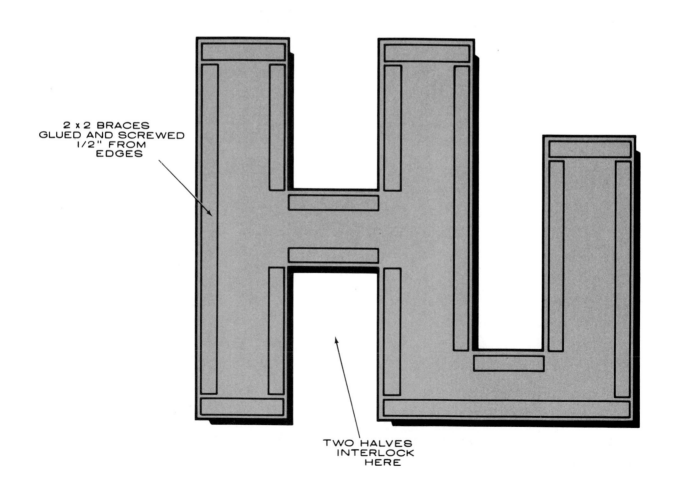

2 x 2 BRACES
GLUED AND SCREWED
1/2" FROM
EDGES

TWO HALVES
INTERLOCK
HERE

ROLLING HOTBED GREENHOUSE

Designed by an avid home gardener to start seedlings well before spring planting, this greenhouse was put on wheels to take full advantage of late-winter solar gain. You can roll it around your deck or patio to chase the drifting sunlight, and wheel it indoors (the garage will do) at night. On cool, sunless days (and during the night) you just plug in the hotbed.

All dimensions given are finished size, so allow for saw kerfs when making your layout on the ¾″ plywood panel. Use waterproof glue and galvanized nails or screws for all joints.

After cutting out all parts (a scroll saw will round the ends), tack the end panels together and drill both at the same time, to take the five dowel rods. Rout ¼″ × ¾″ grooves an inch up from

Philip Reilly
Woodland Hills, CA

the bottom edge of ends and sides and assemble the box, spacing screws about every nine inches. Assemble the leg angles using four screws per leg; notch the top corners as shown, then attach the legs to the box using four more screws per leg.

Stand this assembly on the cross-tie pieces and trace the cutting pattern on them from the leg angles. Cut cross-ties and caster blocks to fit (note that the two caster blocks that go on top the lower cross-tie must be cut to fit *within* the leg angles, like the ends of the upper cross-tie). Glue and screw this assembly between the legs, then add the end braces.

Treat both interior and exterior of the box with preservative. When dry, paint the box exterior (and all other parts) with two coats of a good outdoor enamel.

Cut a length of aluminum angle to serve as an anchor for the plastic sheet, and trim it to fit between notches on the back side of the box. Trim the sheet to the inside box dimension and sandwich the sheet between the rear box side and the angle, letting the sheet extend ¼" beyond, as shown in the detail. Fasten angle and sheet to box side with roundhead screws spaced 12" apart.

Cut the remaining angle and the aluminum bar to the overall length of the box, to be the retractable front edge of the plastic sheet. Drill five holes (3" from ends and equally spaced between) through both bar and upper leg of angle, with these parts slightly offset, as shown in the detail. Pull the plastic snugly over the dowels; sandwich the free end between bar and angle, anchoring with the machine screws. *Note:* A scrap of hardboard glue-nailed atop each leg latches the ends of the angle in place.

PANEL LAYOUT

¾" × 4' × 8'

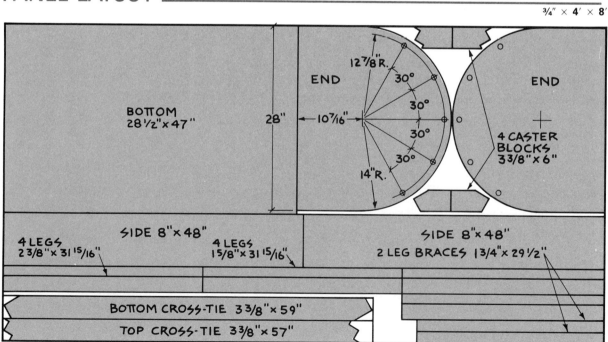

■ MATERIALS LIST

Quantity	Description
1 4 x 8 panel	¾″ plywood, A-C Exterior or MDO
5	Dowels, ¾″ dia. x 4′ for top
1 piece	Clear plastic sheet, 12′ x 4′ x 4 mils
8 ft	Aluminum angle, ¾″ x ¾″ x ¹⁄₁₆″ thick for plastic sheeting anchor and front edge assemblies
4 ft	Aluminum bar, ¾″ x ⅛″ thick for plastic sheeting front edge assembly
6	Wood screws, roundhead, #8 x ⅝″ for plastic sheeting anchor assembly
5	Machine screws, flathead, #4-40 x ⁵⁄₁₆″ with lock washers and nuts for plastic sheeting front edge assembly
8	Box nails, 6d, for caster block assembly
½ gross	Wood screws, flathead, #12 x 1¼″ for box and leg assembly
Small can	Waterproof glue for box assembly
2	Carriage bolts, ¼″ dia. x 2½″ with washers, lock washers, and nuts for crosstie assembly
4	Casters (Sears Cat. No. 9A7410 or equivalent)
1 quart	Wood preservative (Chase Green Copper or equivalent)
2 pieces	Plywood or hardboard scrap, 1⅜″ x ¾″ x ¼″ thick for top of two legs
1 pint	Exterior paint
Small box	Carpet tacks, double pointed for bed preparation
1	Heating cable (Sears Cat. No. 34K4721 or equivalent) for bed preparation
4 ft each	Poultry netting, 30″ wide, hardware cloth, 28″ wide, for bed preparation
1 piece	Fiberglass mat, 1″ thick x 28″ x 4′ for bed preparation
1 roll	Household aluminum foil (heavy duty) for bed preparation
4 cu. ft	Potting soil
2 cu. ft (approx.)	Sand to cover heating cable 2″ deep
1	Extension cord, 25′ (Sears Cat. No. 34A7797 or equivalent)
2	Deck cleats, 6″, with screws, for cord storage

FRONT VIEW

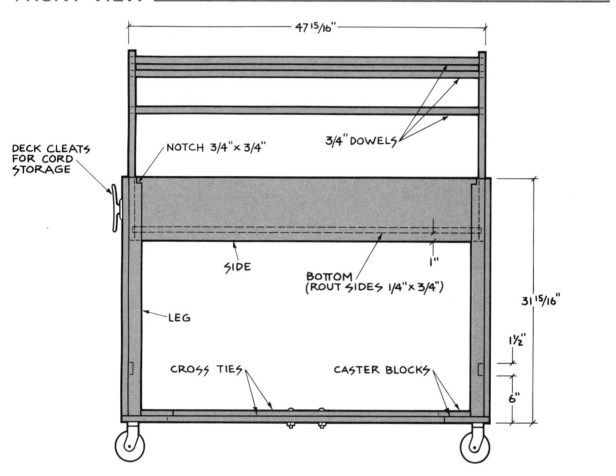

47 15/16"

DECK CLEATS
FOR CORD
STORAGE

NOTCH 3/4" × 3/4"

3/4" DOWELS

SIDE

BOTTOM
(ROUT SIDES 1/4" × 3/4")

1"

LEG

31 15/16"

CROSS TIES

CASTER BLOCKS

1 1/2"

6"

DETAILS

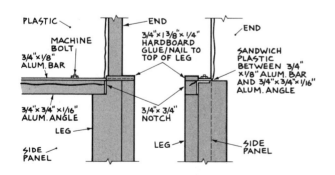

PLASTIC

END

END

MACHINE
BOLT

3/4" × 1 3/8" × 1/4"
HARDBOARD
GLUE/NAIL TO
TOP OF LEG

SANDWICH
PLASTIC
BETWEEN 3/4"
× 1/8" ALUM. BAR
AND 3/4" × 3/4" × 1/16"
ALUM. ANGLE

3/4" × 1/8"
ALUM. BAR

3/4" × 3/4" × 1/16"
ALUM. ANGLE

3/4" × 3/4"
NOTCH

LEG

LEG

SIDE
PANEL

SIDE
PANEL

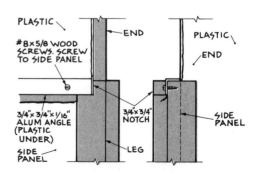

PLASTIC

END

PLASTIC

END

#8 × 5/8 WOOD
SCREWS. SCREW
TO SIDE PANEL

3/4" × 3/4"
NOTCH

3/4" × 3/4" × 1/16"
ALUM ANGLE
(PLASTIC
UNDER)

SIDE
PANEL

LEG

SIDE
PANEL

END VIEW

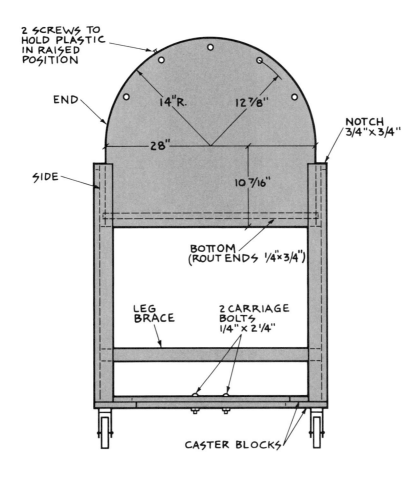

2 SCREWS TO HOLD PLASTIC IN RAISED POSITION

END

14"R.

12 7/8"

NOTCH 3/4"× 3/4"

SIDE

28"

10 7/16"

BOTTOM (ROUT ENDS 1/4"× 3/4")

LEG BRACE

2 CARRIAGE BOLTS 1/4"× 2 1/4"

CASTER BLOCKS

BED DETAIL

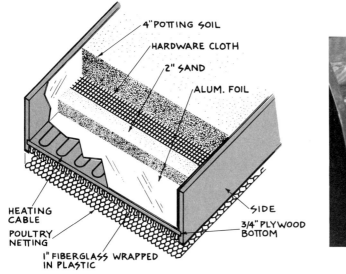

4" POTTING SOIL

HARDWARE CLOTH

2" SAND

ALUM. FOIL

SIDE

3/4" PLYWOOD BOTTOM

HEATING CABLE

POULTRY NETTING

1" FIBERGLASS WRAPPED IN PLASTIC

CROSS BRACE ASSEMBLY

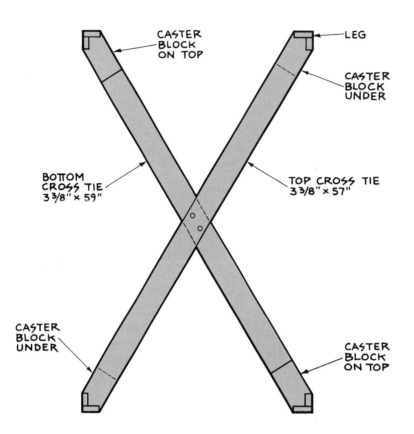

CASTER BLOCK ON TOP

LEG

CASTER BLOCK UNDER

BOTTOM CROSS TIE 3 3/8" × 59"

TOP CROSS TIE 3 3/8" × 57"

CASTER BLOCK UNDER

CASTER BLOCK ON TOP

CROSS BRACES

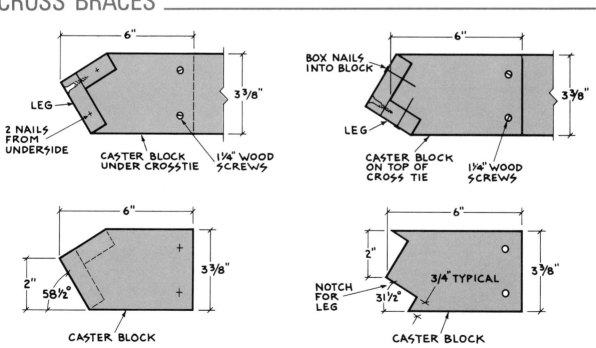

6"

3 3/8"

LEG

2 NAILS FROM UNDERSIDE

CASTER BLOCK UNDER CROSSTIE

1 1/4" WOOD SCREWS

6"

BOX NAILS INTO BLOCK

3 3/8"

LEG

CASTER BLOCK ON TOP OF CROSS TIE

1 1/4" WOOD SCREWS

6"

3 3/8"

2"

58 1/2°

CASTER BLOCK

6"

2"

NOTCH FOR LEG

31 1/2°

3/4" TYPICAL

3 3/8"

CASTER BLOCK

SOLAR DOGHOUSE

Three glazed roof windows, sited due south, admit enough heat from the winter sun to take the chill off the interior of this canine retreat. But since you don't want a hot dog on warm days, the top lifts off so you (and a helper) can rotate it to point the "collectors" north—away from the sun. This removable roof also facilitates cleaning. Being "in the doghouse" was never so good. And all the project requires are 2¼ sheets of exterior plywood: The two full sheets of ½" ply are for the walls and all roof parts; the 2'×4' panel of ¾" ply makes a sturdy, off-the-ground floor for Fido.

You also need lumber for framing and trim, plus Plexiglas for the glazing. Since the base is a simple box with a 12″ × 20″ door

Skip Romero
Jamaica, NY

opening, the only tricky framing is for the gambrel roof. Make a full-size template of a roof end to simplify cutting the 2 × 2s at the proper angles for assembly into rafters.

After you've assembled the three four-sided boxes for the solar windows, paint them white and cut the Plexiglas panels to fit. This is most easily done with a scribing tool you can pick up inexpensively at your plastic dealer's; while there, ask him for a special bit for drilling Plexiglas. Note that the top Plex panel of each window is enough longer than the bottom one to lap it, creating a drip edge that prevents leakage. To stop leaks around the holes, use screws or nails with Neoprene washers under their heads. And be sure the holes are larger in diameter than the shank of the fastener; Plex needs some expansion room.

Since the doghouse emulates an old fashioned barn, we painted our prototype barn red, with a black roof. A good exterior enamel will work on the roof, but for greater authenticity you might want to cover it with strips of asphalt roll roofing.

PANEL LAYOUTS

½″ × 4′ × 8′   ½″ × 4′ × 8′   ¾″ × 2′ × 4′

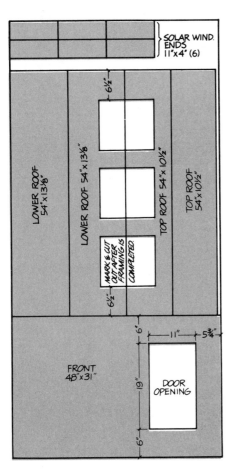

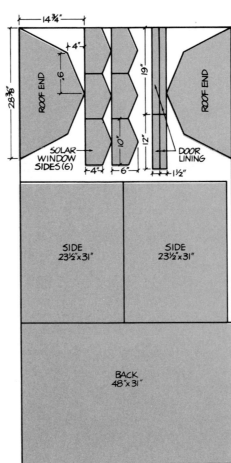

EXPLODED VIEW

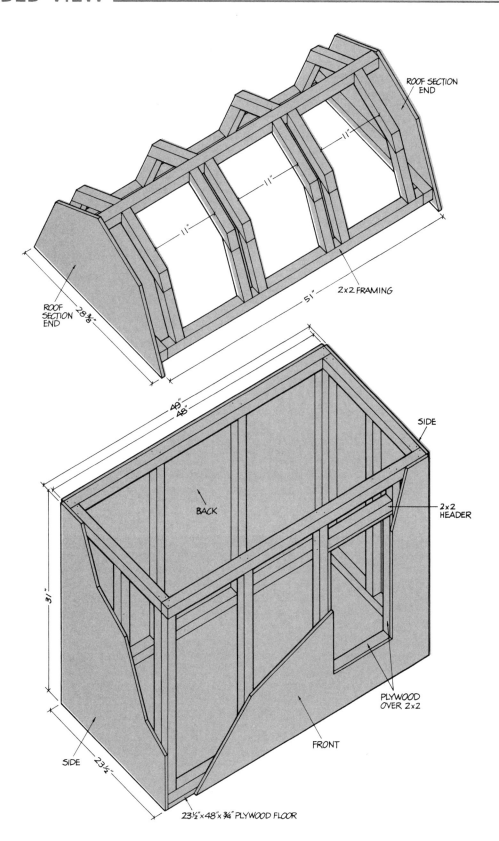

ROOF SECTION
END

11"

11"

11"

2x2 FRAMING

51"

ROOF
SECTION
END

28⅜"

48"
48"

SIDE

BACK

2x2
HEADER

31"

PLYWOOD
OVER 2x2

FRONT

SIDE   23½"

23½"x48"x¾" PLYWOOD FLOOR

MATERIALS LIST

Quantity	Description
2 4 x 8 panels	½″ plywood, MDO or A-B or A-C or B-C Exterior, or APA Rated Sheathing Exterior
1 2 x 4 panel	¾″ plywood, similar grade
14	2″ x 2″ x 8′ lumber framing
4	1″ x 3″ x 8′ lumber trim (for corners and door)
1	2″ x 4″ plexiglass panel and screws (cut to fit)
As required	Asphalt or wood shingle roofing, or as desired, roofing staples or nails, caulking, finishing nails, white or urea resin glue, wood dough or synthetic filler, sandpaper, top quality finish

ROOF DETAILS

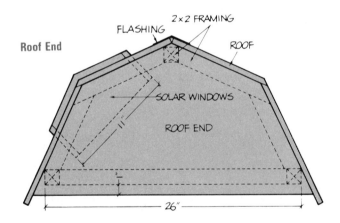

Roof End

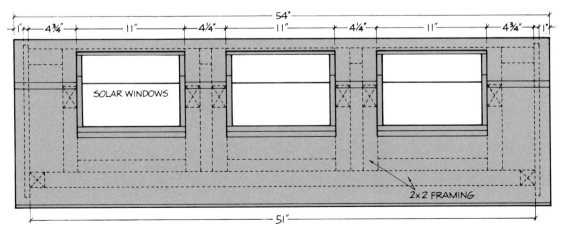

Roof Front

SIDE VIEW & TOP VIEW DETAIL

Top View Detail

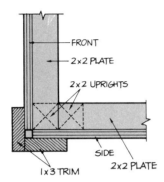

FRONT
2 x 2 PLATE
2 x 2 UPRIGHTS
SIDE
2 x 2 PLATE
1 x 3 TRIM

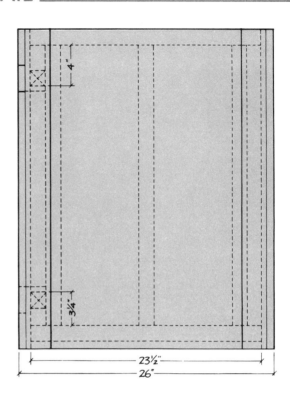

4"

3¼"

23½"

26"

FRONT VIEW

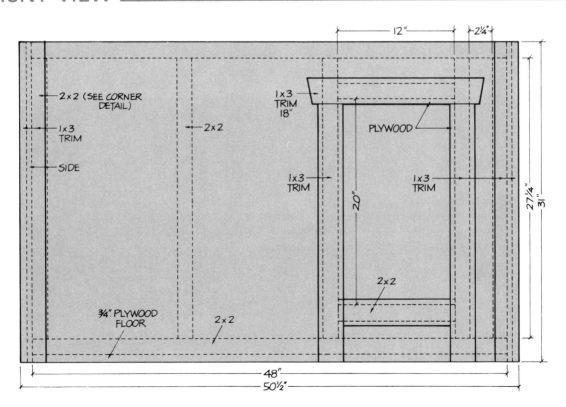

12"

2¼"

2 x 2 (SEE CORNER DETAIL)

1 x 3 TRIM

SIDE

1 x 3 TRIM 18"

PLYWOOD

2 x 2

1 x 3 TRIM

1 x 3 TRIM

20"

27¼"

31"

2 x 2

¾" PLYWOOD FLOOR

2 x 2

48"

50½"

ICE-FISHING HUT

Though it holds two fisherfolk snugly, this unit (designed on the covered wagon principle) quickly knocks down into a plywood box for transport (in any station wagon or van) and for handy storage. When you go back to your snowy fishing site, the cleverly-designed box has sled skids that let you tow the unit from your parking spot to the middle of the lake—using a snowmobile or ordinary man power. When you've reached your fishing hole, you can set the whole shelter up in about five minutes.

Note that measurements shown are finished sizes, so allow for saw kerfs when making your layout on the ½″ plywood panel. The layout is loose enough to permit this easily.

After cutting out all parts, glue the pole-holder pieces together

W.R. Peterson
Okemos, MI

to form a pair of large arcs and a pair of small arcs of three thicknesses of plywood each. Clamp until glue sets, then drill out for ¾" clearance at the three pole positions, as detailed.

Assemble the box parts with glue and nails, as shown in the drawings. Shape the ends of the 2 × 4 skids and glue and screw these to the bottom of the box, driving the screws through the floor and into the skids. Glue and screw the pole holders in position at each end. (For the prototype, the pole holders and skids were painted separately for a two-tone effect once they were screwed in place.) After painting the box, attach the hinged seat lids.

Splice each pair of PVC pipe pieces with the 4" conduit and stove bolts (see detail). These pairs disassemble for storage in the box, held by notches in the seat lids.

To create the tent, you can buy canvas duck or any other waterproof material and sew it up with a front access zipper, as shown in a sketch. If you lack sewing skills, you can have this customized by any tentmaker. Add ties to the inside bottom edge approximately 2" above ground level, for fastening to the poles. Once the cover is anchored and you're zipped inside, let the winter gales howl. You'll keep a snug watch over your ice hole—especially if you've brought a portable heater.

Note the seat bottoms are recessed so you can add 2"-thick foam padding and cover it with upholstery material. The "cutout" between these seats becomes the entry threshold, making it easy to step through the zipper-slit while the box is resting on its skids.

PANEL LAYOUT

½" × **4'** × **8'**

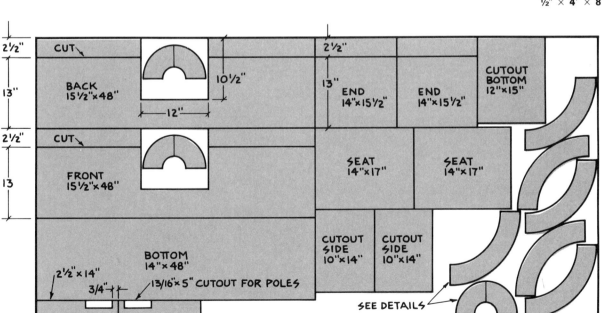

MATERIALS LIST

Quantity	Description
1 4 x 8 panel	½" plywood, A-C Exterior or MDO
2 pieces	2 x 4 lumber, 51" long, for runners
8 pieces	1 x 1 lumber; (4) 7¼" long, (4) 17" long, for seat ledgers
6 sections	Conduit, ¾" (O. D.) x 46½" long for frame
6 sections	PVC pipe, ⅝" (O. D.) x 42" long for frame
3 sections	Conduit, ¾" (O. D.) x 4" long for frame splices
6	Wood screws, #6 x 1½" long for attaching runners
6	Stove bolts, ³⁄₁₆" x 1", with nuts, for PVC/conduit splice
½ lb (approx)	Finishing nails, 6d
1 can (8 oz approx)	Waterproof glue for glue-fastening all joints
As required	Wood dough for filling countersunk nail holes and plywood edge voids; fine sandpaper for smoothing plywood cut edges and filler
2 pieces	Continuous hinge, 12" long, with screws
As required	Exterior primer and paint.
2 pieces	Foam rubber, 13" x 16" x 2" thick for seats
2 pieces (optional)	Cover material for foam rubber seats (cut to fit)
1 piece	Tent material (lightweight canvas, vinyl, nylon, etc.) 5' x 7' for front (cut to fit)
2 pieces	Tent material, 6' x 16' (cut to fit)
1 (optional)	Zipper, brass, 6' long

TOP VIEW

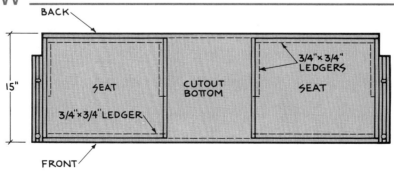

FRONT VIEW

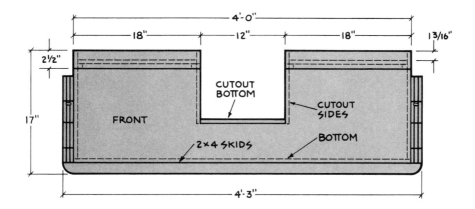

END VIEW

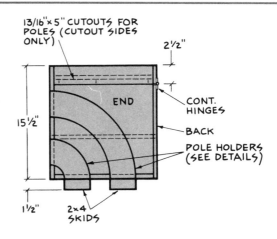

PIPE HOLDERS: DIMENSION DETAIL

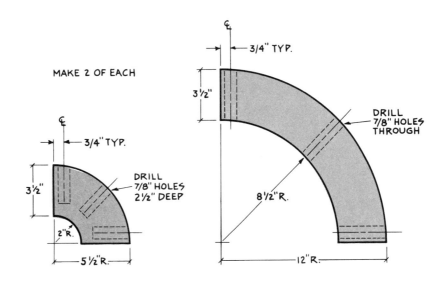

EXPLODED VIEW

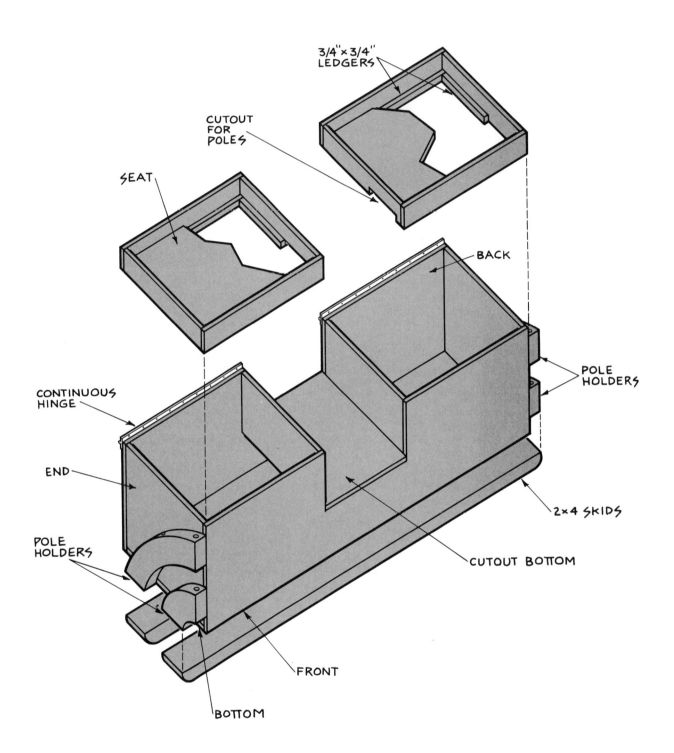

3/4" × 3/4" LEDGERS

CUTOUT FOR POLES

SEAT

BACK

CONTINUOUS HINGE

POLE HOLDERS

END

POLE HOLDERS

2×4 SKIDS

CUTOUT BOTTOM

FRONT

BOTTOM

FRAME ASSEMBLY

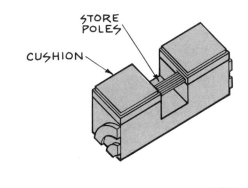

STORE POLES

CUSHION

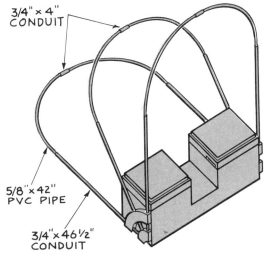

3/4" × 4" CONDUIT

5/8" × 42" PVC PIPE

3/4" × 46½" CONDUIT

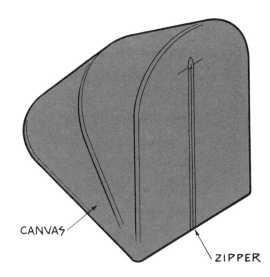

CANVAS

ZIPPER

PIPE ASSEMBLY DETAIL

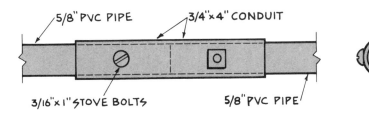

5/8" PVC PIPE

3/4" × 4" CONDUIT

3/16" × 1" STOVE BOLTS

5/8" PVC PIPE

CAR-CARE CART

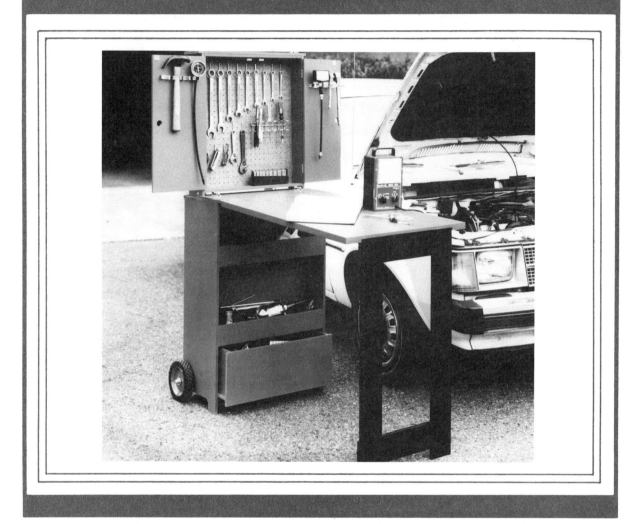

Designed for the driveway mechanic, this unit is a combination tool chest and workbench that folds up and telescopes into a handy cart you can roll into the garage when your tuneup is done.

It all comes out of 1½ sheets of ½″ plywood—and in the original design the minimal waste was cleverly used for a variety of tool holders. In our rebuild, above, we simplified tool-hanging by installing a perf-board panel, spaced out from the back to take commercial hooks and racks. The original designer cut one 42″ length of 1″ piano hinge in two for attaching both the table and the leg unit. The cart has a bottom drawer with a bin above it; these store heavy tools and test equipment, maintaining a low center of gravity that makes for easy wheeling.

Mark R. McClanahan
Clio, MI

Note how cleverly the barrel-bolt latches secure the telescoping tool-box in both its raised and lowered positions. For the former, they shoot forward under the lip of the bottom, to prevent retracting. When the unit is lowered into its pocket, they shoot into their own barrels, mounted on top.

Even the drawer is a simple butt-joint box, with the bottom protruding at each side to ride atop a cleat on the inside face of each side of the cart. You can use threaded rod for the axle, bolting it through the cart as shown, spacing the wheels from the sides with washers. Or use a conventional axle rod drilled for cotter pins, instead.

You'll find the swing-up 24″ × 36″ table top a real boon next to your raised hood—as a stand for test equipment and the auto-

PANEL LAYOUTS

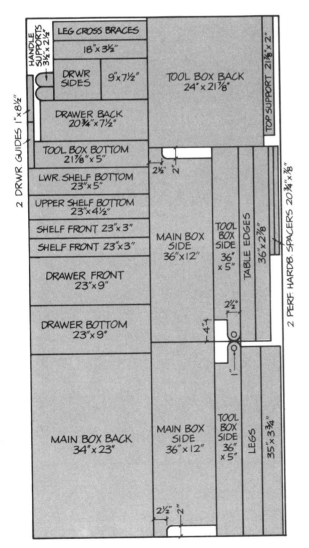

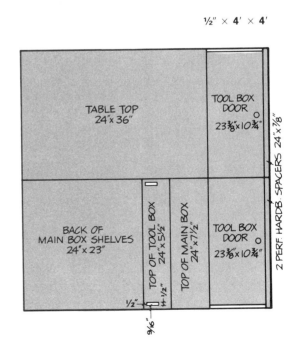

MATERIALS LIST

Quantity	Description
1 4 x 8 panel	½″ plywood, MDO or A-B or A-C Exterior
1 4 x 4 panel	½″ plywood, similar grade
1	1″ x 23″ wood dowel
2	Approx. 1″ wide x approx. 6-8″ dia. wheels with axle, 4 nuts and 6 washers (drill axle holes in plywood sides according to wheel size)
2	1¹⁄₁₆″ x 18″ continuous hinges w/screws (for attaching tabletop)
4	1″ x 1½″ butt hinges w/screws (for attaching toolbox doors)
2	Barrel bolt latches w/screws (for securing toolbox)
2	Magnetic catches w/screws (for toolbox doors)
1	24″ x 21⅞″ perforated hardboard with miscellenous hanger handware as desired
As required	Wood screws, finishing nails, white or urea resin glue, wood dough or synthetic filler, fine sandpaper, top quality finish

maintenance magazine or book you're using as guidance. Though our plans don't call for it, you may want to add a folding leg brace to prevent accidental collapse of the table.

The 1″ dowel provides both the lift handle for the telescoping tool-box and the handle for wheeling the cart in and out of your garage. In its folded mode it easily stores alongside the car, since it requires less than a 16″ depth.

Start construction by ripping your full sheet into two long panels with a portable circular saw. (But since dimensions given are actual, be sure to provide for this saw kerf.) Complete your parts layout on both these panels and separate the remaining parts with your circular saw and a saber saw. Note that the drilled top tabs of the tool-box side pass through slots in the top piece. After the dowel is inserted through one of these holes, it's locked in place by the handle-support "doublers" glued to each outside face. We painted our prototype brick red with black accents—the handle and table leg.

ISOMETRIC VIEW

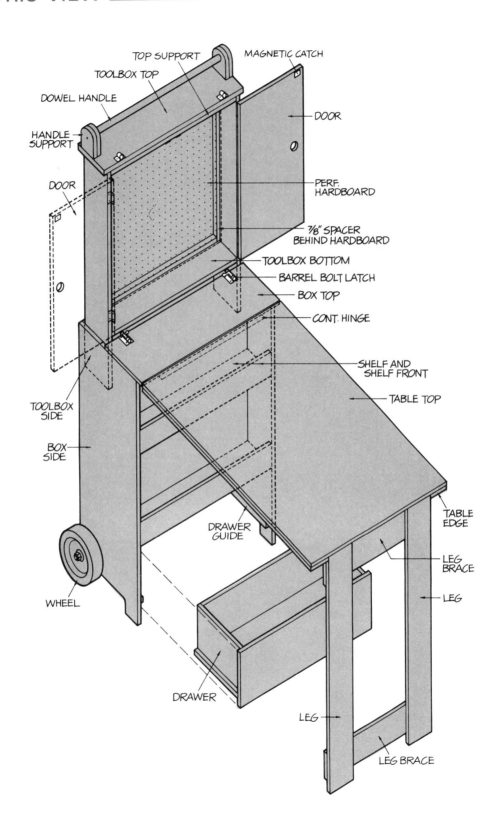

TOP SUPPORT

TOOLBOX TOP

MAGNETIC CATCH

DOWEL HANDLE

DOOR

HANDLE
SUPPORT

DOOR

PERF.
HARDBOARD

7/8" SPACER
BEHIND HARDBOARD

TOOLBOX BOTTOM

BARREL BOLT LATCH

BOX TOP

CONT. HINGE

SHELF AND
SHELF FRONT

TOOLBOX
SIDE

TABLE TOP

BOX
SIDE

TABLE
EDGE

DRAWER
GUIDE

LEG
BRACE

LEG

WHEEL

DRAWER

LEG

LEG BRACE

FRONT & SIDE VIEWS

Front · Side

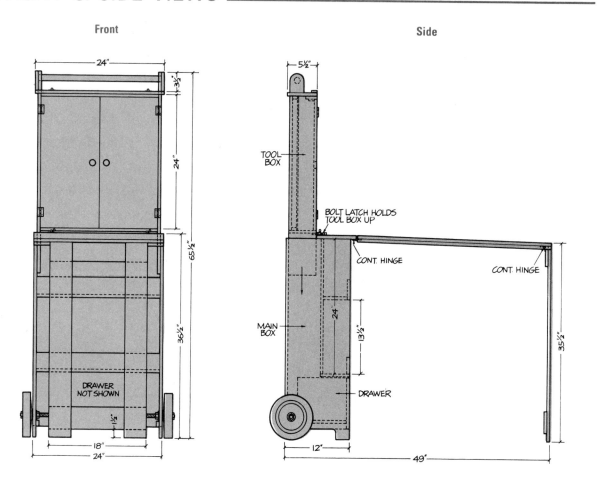

DRAWER DETAIL

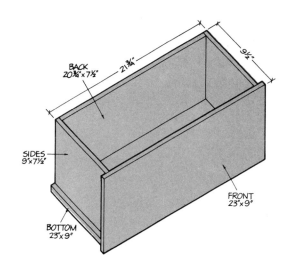

CANOPY FOR PICKUP TRUCK

The weatherproof cap for this Dodge pickup can be adapted to most compact trucks. The ¼"-plywood shell is coated with fiberglass, like a boat hull, and the hinged back lifts for access. Good looking as it is, it's inexpensive to build. Trussed framing and the strength of plywood (two ¼" panels plus a ½" one) make the unit tough enough to endure years of road use, offering watertight protection to whatever you're hauling.

Note that the American Plywood Association draftsman added another half-sheet of ½" plywood for six pairs of 2"-wide legs, for the unit's ribs. Such inefficient use of a plywood panel would have disqualified this entry; the original layout showed many of these parts cut from waste areas in the full-size ½" panel, with scrap pieces from the woodbox used for the balance.

Ken Hurst
Naperville, IL

323

PANEL LAYOUTS

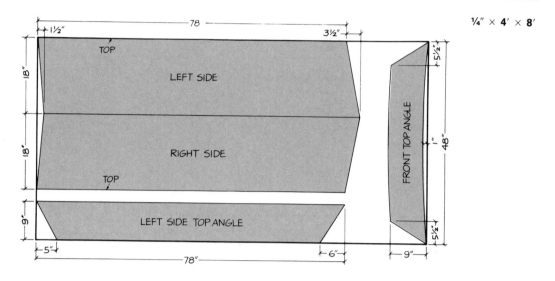

¼″ × **4′** × **8′**

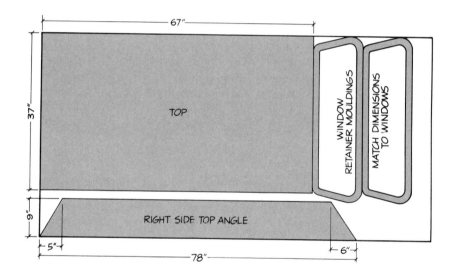

¼″ × **4′** × **8′**

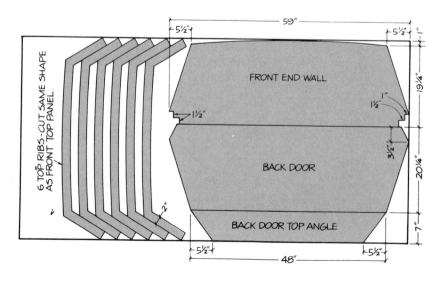

½″ × **4′** × **8′**

PANEL LAYOUT

$\frac{1}{2}'' \times$ **4**$'$ $\times$ **4**$'$

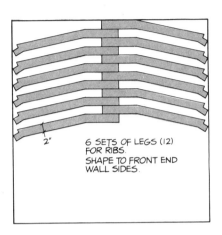

2″

6 SETS OF LEGS (12)
FOR RIBS.
SHAPE TO FRONT END
WALL SIDES.

MATERIALS LIST

Quantity	Description
2 4 x 8 panels	¼″ plywood, A-C Exterior Group 1
1 4 x 8 panel	½″ plywood, same grade
1 4 x 4 panel	½″ plywood, same grade
2	1 x 1 framing lumber (8′ lengths)
3	1 x 2 framing lumber (8′ lengths)
1	1 x 4 framing lumber (8′ length)
6	2 x 2 framing lumber (8′ lengths)
3	2 x 3 framing lumber (8′ lengths)
2	2 x 4 framing lumber (8′ lengths)
2	12″ x 42″ framed sliding windows with screens
2	12″ x 42″ safety glass windows, cut to size or salvaged from wrecking yard
1 pr.	3″ butt hinges with screws
1	"T" handle lock assembly
12	¼″ x 2½″ carriage bolts with nuts and washers
10	⁵⁄₁₆″ x 3″ lag screws with washers
24	⁵⁄₁₆″ x 2½″ lag screws with washers

Begin by laying out your three full panels. Cut out for the front, side, and back windows (bought from an auto-supply store or wrecking yard). Rout a ½″-wide rabbet, as deep as the glass is thick, in the front and back panels, if you are using fixed, salvaged panes here.

Cut the base plates, longitudinal stringers, rib leg spacers and door frame pieces to size. Bevel the plates and stringers to conform to the desired slope of the sides. Positioning of the horizontal roof ribs is optional, depending on the truck-bed dimensions and window placement—so stringer and rib leg spacer lengths will vary.

Laminate roof ribs and rib legs with panel adhesive and carriage bolts. Assemble the skeletal frame as shown. Use diagonal lag screws to attach rib legs through plate stringers to base plates. Use vertical lag screws to attach plate stringers to base plates. Attach the door hinges to the top of the 2×4 frame. (Some shaping of this door frame—and the door—is required to insure a weathertight fit.)

Now attach the plywood skin to the frame, beginning with the front end wall and front wall cap, using panel adhesive and ¾″ #8 wood screws for joints to plywood, and 1½″ #8 wood screws for connecting to framing. Bevel mating plywood edges to aid fit and weathertightness.

Cut door parts to size, assembling as shown, with wood screws and panel adhesive. Sand the canopy and door and fill any gaps with auto body putty. Apply fiberglass cloth and resin, following recommendations with the kit. Sand and prime. Install front and door windows with plywood retainer moldings, wood screws and caulk. Install and caulk slider units in the side panels.

Apply finish coat of enamel and install door. No provisions are shown for hardware to hold door open, but either a folding or collapsing RV canopy brace will do the job.

SIDE VIEW

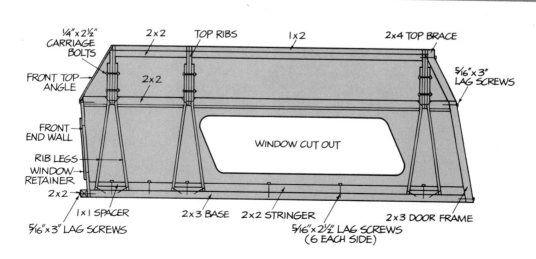

CANOPY: FRONT-VIEW CUTAWAY

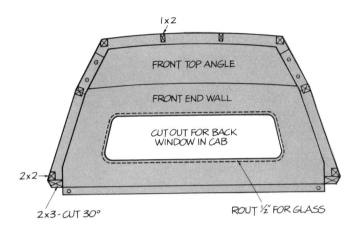

1x2

FRONT TOP ANGLE

FRONT END WALL

CUT OUT FOR BACK
WINDOW IN CAB

2x2

2x3 - CUT 30°

ROUT ½" FOR GLASS

CANOPY BACK

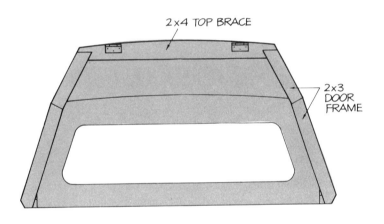

2x4 TOP BRACE

2x3
DOOR
FRAME

CANOPY: BACK-VIEW DOOR DETAIL

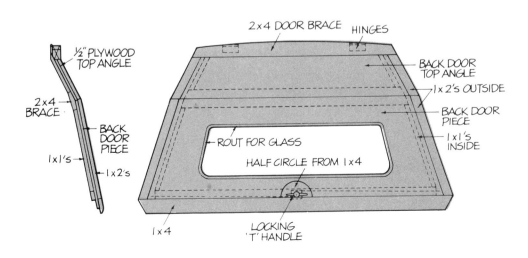

½" PLYWOOD
TOP ANGLE

2x4
BRACE

BACK
DOOR
PIECE

1x1's

1x2's

2x4 DOOR BRACE   HINGES

BACK DOOR
TOP ANGLE

1x2's OUTSIDE

BACK DOOR
PIECE

1x1's
INSIDE

ROUT FOR GLASS

HALF CIRCLE FROM 1x4

1x4

LOCKING
'T' HANDLE

FLAT-ENDED DINGHY

This minimal boat is built from two different-thickness sheets of plywood, plus a 1 × 4 and 1 × 2s. The latter, spaced on cleats, form a duckboard floor to keep your feet dry. You cut the paddles for the oars from the ½" sheet and use 1½"-dia. poles for the handles. The ¼" marine-grade plywood is cut only once, to form the bottom and the prow. Flat-bottomed and wide for stability, this little boat is just the right size for an angler to sneak into those hidden coves where the big ones lurk. The seat has a compartment underneath for stowing a tackle box and, if you wish, you can insert flotation material (such as Styrofoam block) into the end compartments.

The boat's compact enough for easy transport, and you don't need any special tools to construct it, though a portable circular saw will speed cut-out of the parts, and a saber saw will quickly shape the necks of the paddles. No provision has been made for

William Schroeder
Merritt, BC, Canada

saw kerfs in the layout dimensions, so center your cuts on the pencil lines. Lightly sand all cut edges for best glue contact.

Construct the boat frame from 1×2 lumber as shown in the exploded view (and side bow detail). Use waterproof glue and 3d nails or corrugated fasteners to make strong joints. Don't install floor boards or the middle three bow supports at this time.

PANEL LAYOUTS

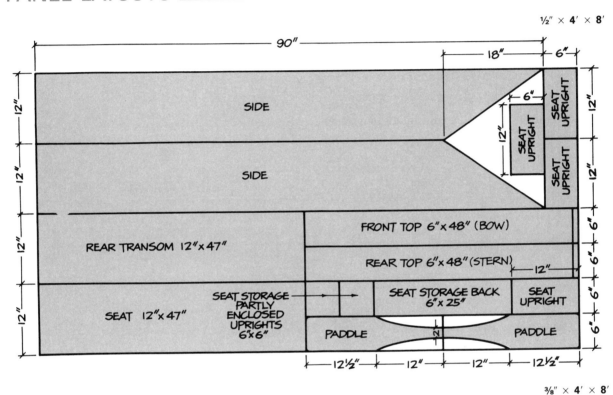

$\frac{1}{2}'' \times 4' \times 8'$

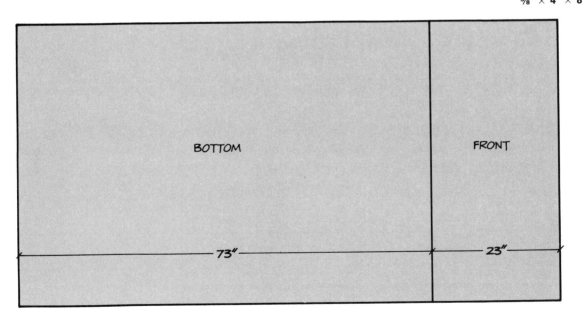

$\frac{3}{8}'' \times 4' \times 8'$

MATERIALS LIST

⅜" ½"

Quantity	Description
1 4 x 8 panel	⅜" plywood, Marine, A-B Exterior, or MDO
1 4 x 8 panel	½" plywood, same grade
As follows:	1 x 2 lumber:
6 pieces	9¾" long for vertical framing
3 pieces	45½" long for top cross framing
1 piece	44" long for transom base framing
2 pieces	88¼" long for side top framing
2 pieces	72¾" long for side bottom framing
3 pieces	67¾" long for bottom panel nailers
5 pieces	21⅜" long for front (bow) supports
22 pieces	47" long for floor cross strips
2 pieces	78¾" long for side top trim
1 piece	72¼" long for keel
1 piece	1 x 4 lumber 44" long for spreader
2 each	Commercially available oarlocks and mounting sockets (obtain from marine supply outlet)
4 each	¼" x 3" long carriage bolts with washers and nuts for oarlock mounting block assemblies
2 pieces	Wood block approx. 1" x 2⅜" x 5" for oarlock mounting block assemblies
2 pieces	Wood block approx. 1" x 2¾" x 5" for oarlock mounting block assemblies
2 pieces	1½" dia. x 57" long dowel for oar handles
As required	Waterproof glue
As required	2d and 3d non-staining, non-corrosive casing or finishing nails for glue-nailing
20 each	1½" corrugated fasteners for glue-fastening
60 to 70 each	#8 x 1½" brass or stainless steel, oval- or flathead wood screws
8 each	#8 x ¾" brass or stainless steel, oval-or flathead wood screws
As required	Waterproof wood preservative compatible with finish; fine sandpaper for smoothing plywood cut edges; wood-compatible marine paint for finishing

Lay the plywood bottom panel down, bottom face up and set the frame, right side up, on top. Locate the frame ½" in from each side and ½" forward of the aft, just as it will later be fastened to the reverse side of the panel. Trace pencil outlines of the bottom framing pieces. Remove the frame and set it bottom side up. Support the bottom frame members with blocks so you'll be able to drive nails into them through the bottom panel. Coat the bottom surfaces of the frame pieces with waterproof glue and set the plywood panel in place, bottom face up, exactly as before. Using 2d finishing nails, fasten the panel to the frame by driving within the penciled guidelines, every 6" along all framing members.

When the glue has set, turn the boat right side up and install the transom, using glue and 1½" screws every 12" (drill pilot holes first). Next, install the boat sides the same way.

Bevel the edges of the plywood bow piece as necessary, drill pilot holes and install with glue and 1½" screws. You can bevel the front end of the bottom panel to match the bow at this time, with a block plane and a wood rasp.

Install the three remaining bow supports and floorboards (except those for the seat base) as shown in the exploded view. For bow supports, use glue and 1½" screws through the bow piece.

EXPLODED VIEW

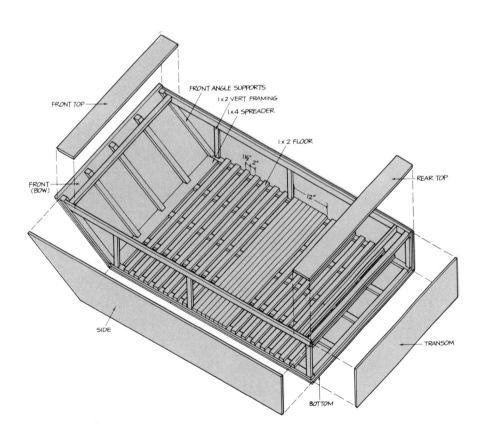

Start the seat assembly outside of the boat by edge-gluing eight lumber strips together as a seat base. After drilling pilot holes, use 1½″ screws (and glue) to fasten the middle four seat supports and back to the seat base and each other. Don't install the two outermost supports. Install this assembly in the boat, glue-nailing the seat base to the bottom 1 × 2 nailers. Drill pilot holes and use glue and four ¾″ screws at each side to install the outer seat supports against the boat sides, driving the screws through the boat sides. Glue-screw the seat in place.

After installing gunwales, oaklock blocks and 1 × 2 keel as shown, brush on a good marine enamel, thoroughly coating all joints. The oars, assembled with waterproof glue and screws, can be painted to match, then set into purchased oaklock sockets.

SIDE BOW DETAIL

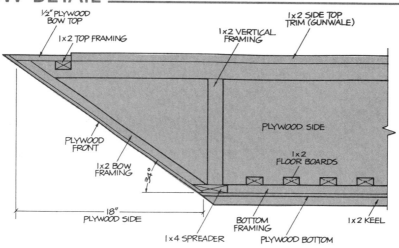

SEAT ASSEMBLY & OARLOCK MOUNTING ASSEMBLY

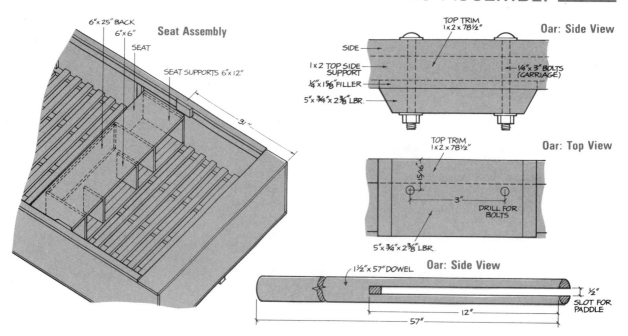

PORTABLE DINGHY

Strictly a one-man boat, this neat little shell can be lifted with one hand, and is compact enough to be carried on the roof of a car or slipped over the tailgate of a station wagon. Depending on your epoxy coating, the ¼"-plywood hull will weigh in at 25 to 30 pounds. Note that our designer met the contest strictures against waste by cutting two oar blades from corners of the 4' × 8' panel. (To simplify your layout, mark the whole panel off into a 6" grid, as shown.)

Cut out all pieces with care, using a saber saw with a fine-toothed blade. Clamp the two side and bottom pairs together and trim each pair with a sharp hand plane to insure symmetrical perimeters on each side of the keel. Bevel butting edges.

James Richardson
Boston, MA

Drill small holes on 6″ centers along mating edges and temporarily "sew" the six panels together with twists of copper wire. Brace the parts to spread the gunwales to a 40″ beam. Apply small beads of expoxy along all joints, inside and out. When this has cured, remove the copper stitches.

Buy epoxy/microballoon joint compound and mix thickly, according to label directions. Apply this to produce a fillet along all interior seams, as shown in the detail sketches. Cut the stern and bow transom reinforcements from a 1 × 10; chamfer the edges as shown and glue these in place. Bevel 1 × 1s to form gunwale strips, cut to length, allowing for curvatures, and glue in place, with clamps. Peg the gunwale ends with dowels as shown in the main assembly drawing.

Rip the seat from the 1 × 10 and trim it to beam width (about

PANEL LAYOUT

¼″ × 4′ × 8′

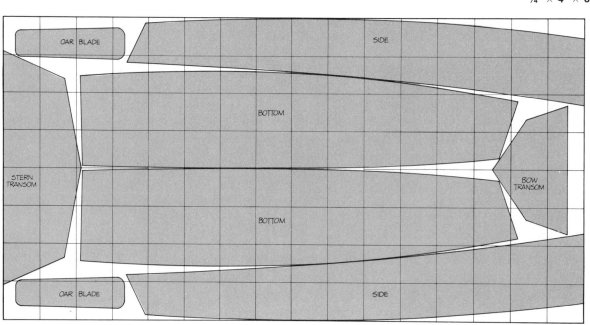

MATERIALS LIST

$\frac{1}{4}''$

Quantity	Description
1 4 x 8 panel	¼″ plywood, Marine A-A Exterior
8 lin. ft	1 x 10 lumber
24 lin. in.	1 x 1 framing lumber
12 lin. ft	1⅝″ dia. closet rail
24 lin. in.	⅝″ dia. hardwood doweling
12 lin. in.	¼″ dia. hardwood doweling
1 pr.	Oarlocks
4	Flathead wood screws
As required	16-gauge copper wire, scrap oak for oarlock blocks, epoxy and microballoon joint sealing compound (available at yacht chandlery), scrap leather, copper brads, sandpaper, top-quality marine-grade enamel paint and companion primer

ISOMETRIC VIEW

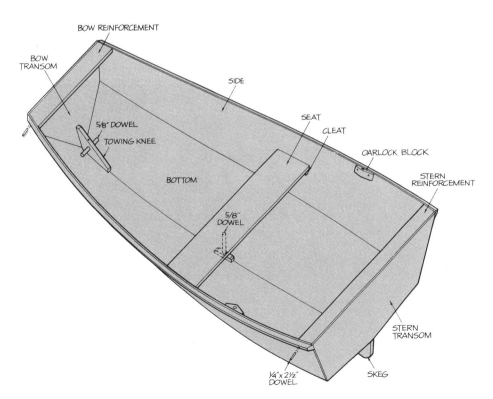

34″ on center from stern transom and 5½″ above bottom center seam. Chamfer the underside as shown in a detail and shape the seat ends to the side curvatures. Cut the seat cleats, dowel support and dowel-support base, shaping as shown. Drill a socket in both the seat bottom and the base and glue all parts in place.

Cut out the towing knee (dimensions and final shape not critical). Drill for and install the dowel, then drill a hole for the tow rope right through the bow transom. Reinforce all these joints with filleting similar to that applied earlier.

Turn the boat over to trim and sand all exterior seams. Fill all gaps with an epoxy bead. Suggested dimensions for the skeg are 24″ long by about 4″ deep at the stern, tapering to 1″ at the narrow end. Round the corners, shape it to fit the hull and glue in place, reinforcing joints with filleting.

Prepare inside and outside for painting; apply at least two coats of a top-quality marine-grade enamel paint, wet-sanding between coats. Check detail for construction of the oars. You taper the 6′ pole from the oaklock position (about 18″ from hand-grip end) to ¾″ at the blade end. Shape the handle octagonally for best gripping surface. Glue the blades into slots and reinforce joints with filleting.

ASSEMBLY DETAILS

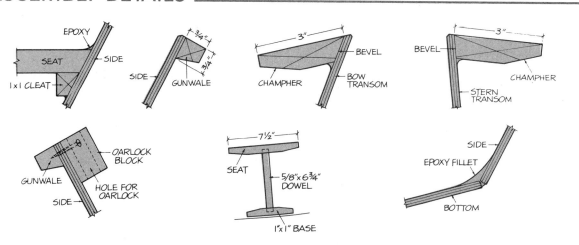

OAR ASSEMBLY & SKEG DIMENSIONS

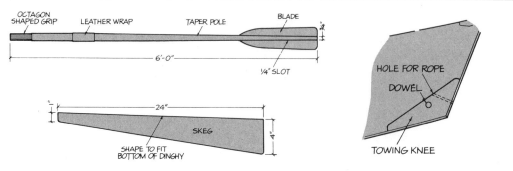

SPLIT-APART DINGHY

This dinghy separates into independently-floating halves, nests inside a hatchback or small station wagon, *and* can take a trolling motor! Piano hinges across the bottom and sides and ¼″ ply are the keys to its versatility.

The designer, a tool-and-die maker, wanted a lightweight boat to replace his inflatable. "Inflatables are for the young," he told me, "and as my wife and I grew older, we found it harder to climb in and out of ours. I didn't want to wrestle a boat onto the top of my hatchback and, since we tow a travel trailer, the only way we could take a boat along was to tuck one into the back of the car, as we'd done with the inflatable."

"Loose-pin piano hinges proved an ideal way to split my design

Frank Distler
Cold Spring, KY

PANEL LAYOUTS

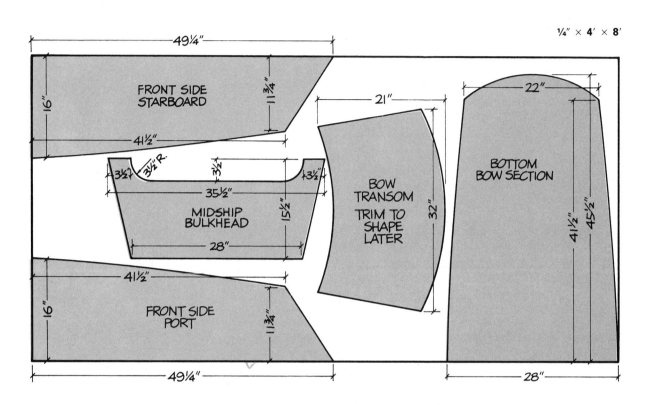

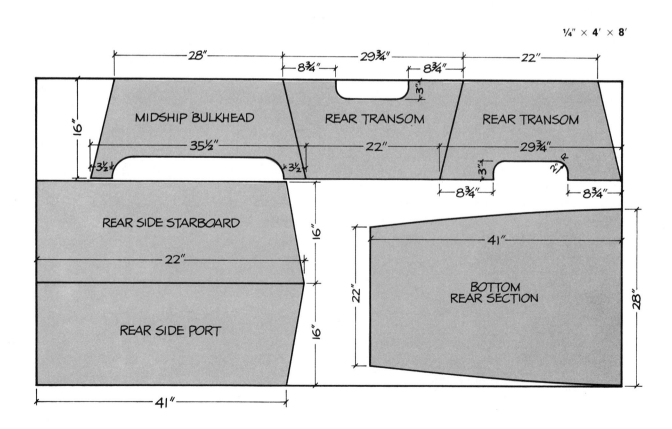

◾ MATERIALS LIST ▰

Quantity	Description
2 4 x 8 panels	¼″ plywood, Marine A-A Exterior, if available, or A-C Exterior Group 1
1	5′ x 64″ fiberglass cloth (cut into 2½″-wide strips and applied as shown)
2 quarts	Polyester resin
App. 110	¾″-wide x 1¹³⁄₁₆″-long aluminum angles (cut from .050″-thick scrap aluminum and bent to the required angle)
App. 220	⁵⁄₃₂″ dia. x ⅜″-long flathead tubular soft iron rivets
2	36″ continuous steel hinges with undersized wire loose pins (for joining pram sections)
1	24″ continuous steel hinge with undersized wire loose pins (for removable front seat)
App. 36 lin. ft.	1 x 2 lumber framing (for keel, gunwale, bulkhead and transom)
As required	Wood dough or synthetic filler, sandpaper, wood screws, top quality marine paint

into nesting halves. I wanted to keep the weight down, so I went with ¼″ ply and omitted framing; I doubled up the transom, though, to make it sturdy enough for a small motor. I rub a little Vaseline on the wire pins each time I feed them through the mated hinge barrels.''

But how do you join such thin panels? Here's the machinist's triumph: He *riveted* them all together, using aluminum angles, as shown in the section detail. You can't get much more lightweight than that! The assembled dinghy is eight feet long by three feet wide and weighs under 62 pounds. For nesting the halves, of course, the front seat (which can be ⅝″ ply or 1 × 10 lumber) must be lifted off its cleats. The only lumber piece that's permanently installed in the front section (aside from the seat cleats and bulkhead framing) is the front deck, which was saber-sawed to shape from ⅝″ white pine and glued and screwed in place. (The framing at the bulkhead provides a firm connection for the hinge screws.)

The .05″-thick aluminum was sheared in ¾″-wide strips and then snipped into 1³⁄₁₆″ lengths. You bend each piece to match the angle for its location, then drill right through aluminum and plywood to create the rivet holes. Note in the ''bottoms-up''

SIDE VIEW

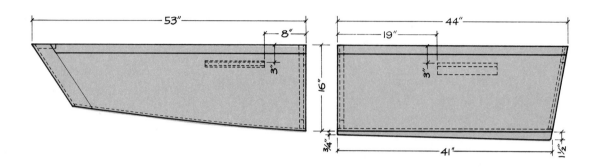

ISOMETRIC VIEW

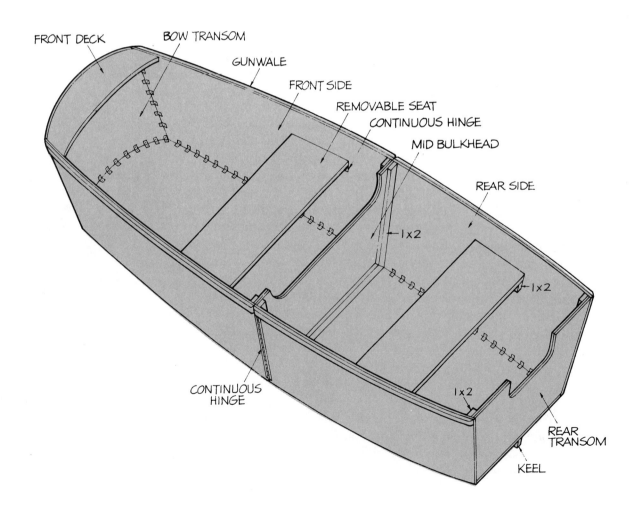

BULKHEAD & TRANSOM DIMENSIONS

Mid Bulkhead   **Stern Transom**

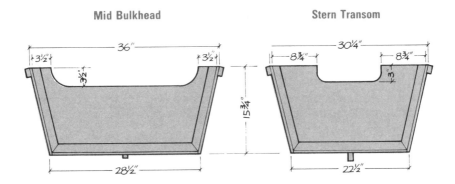

ISOMETRIC VIEW: BOTTOM WITH HINGE

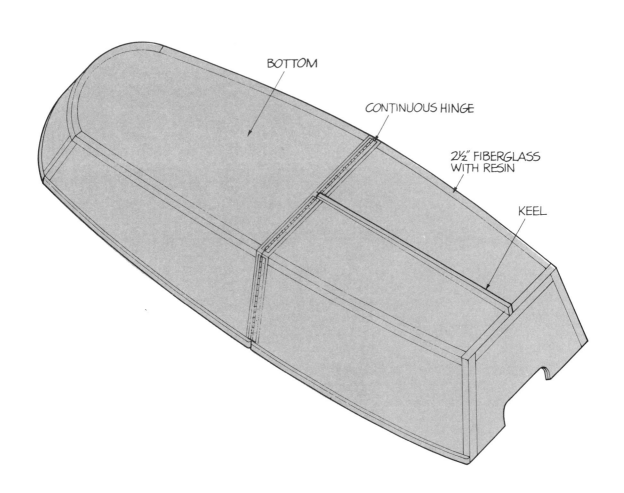

sketch, the detail, and a photo, how all joints are round and capped with fiberglass strips laid in polyester resin. You can buy both fiberglass and resin—with instructions—at any marine supply store. Treat the inside of all joints the same way, to cover the riveted angles. In the prototype, the piano hinges were applied *after* the bulkhead joints were taped. To insure no leaks, you may prefer to apply the hinges first and do the exterior taping right over the screwed-on leaves.

Attach a ½″ × ¾″ rub rail with resin and screws after all fiberglassing is complete. Brush on a good marine primer and compatible top coat.

ASSEMBLY DETAIL

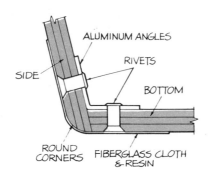